PRAISE FOR *RESTORING THE PITCHFORK RANCH*

"This book will help spur the imagination of other landowners: 'how can I help?' is the most human of questions, and it turns out the answers are manifold!"

—Bill McKibben, author of *The End of Nature*

"A. T. Cole is as great of a writer as he is a practitioner of stewarding rangelands and their community. In a region currently being devastated by drought, wildfires, and political divisiveness, Tom is not restoring valuable relationships but re-storying the way we relate to the land. Bravo!"

—Gary Paul Nabhan, co-author of *Agave Spirits: The Past, Present and Future of Mezcals*

"The great American conservationist Aldo Leopold once wrote that 'One of the penalties of an ecological education is that one lives alone in a world of wounds.' Tom and Cinda Cole learned this wisdom firsthand when they purchased the Pitchfork Ranch. Decades of hard use had created a variety of ecological wounds—some easy to recognize, others only revealed as the Coles came to know their land. In the finest Leopold tradition, they set out to heal these wounds and make the land healthy again, which will be increasingly important under climate change. It is a story for our times—and it's an inspiring one!"

—Courtney White, author of *Grass, Soil, Hope: A Journey through Carbon Country*

"A riveting tale that combines history, advocacy, and how-to, *Restoring the Pitchfork Ranch* is both a kick-in-the-butt call for individual action on climate change and an inspiring story of what one couple with a passion for restoring the land can accomplish."

—Susan J. Tweit, author of *Bless the Birds: Living with Love in a Time of Dying*

"A. T. Cole's book has arrived in desperate times. Cole gives us a road map out of the morass of climate change, species extinctions, and catastrophic soil loss. Using their Pitchfork Ranch as a learning lab, he and Lucinda, his wife, demonstrate how restoring lands and waters can address this trifecta, whether you are urban, suburban, or rural. Critically, it gives us hope for a future where people and land get on better together."

—Richard L. Knight, Colorado State University

"A. T. Cole's story of land restoration on the Pitchfork Ranch rests on a firm premise and a promise: that however daunting the world's multiple and intersecting crises may be, all of us can be agents of positive change. In this corner of New Mexico's high desert grasslands, Cole and his wife, Lucinda, have devoted themselves to repairing a wounded place. In sharing the story of their work, Cole encourages us all to take up our part in healing a wounded world."

—Curt Meine, author of *Aldo Leopold: His Life and Work*

"For the past 140 years, Arizona and New Mexico's precious ciénagas have been extensively drained, degraded, and destroyed. Twenty years ago, at Burro Ciénaga, the Coles reversed the process. After removing the cattle and installing hundreds of stream stabilization structures, gully erosion has ceased and recovery is now well underway. Much of the ciénaga has been re-wetted. Wetland vegetation and wildlife have returned. My favorite rule applies: When it stops getting worse, it starts getting better. Well done, you two!"

—Bill Zeedyk, co-author of *Let the Water Do the Work: Induced Meandering, an Evolving Method for Restoring Incised Channels*

"*Restoring the Pitchfork Ranch* is not about trying to put an arid landscape back the way it once was. It's about water retention, soil building, biodiversity conservation, and carbon sequestration. All this alongside responsible cattle grazing. Passionately written, it's ultimately about hope and how together we can overcome the intersecting environmental crises that imperil our very existence on Planet Earth."

—J. Baird Callicott, author of *Thinking Like a Planet: The Land Ethic and the Earth Ethic*

"There are 770 million acres of rangeland in the United States, much of it desertified through watershed destruction and overgrazing. In a masterfully written account of the restoration of the Pitchfork Ranch, A. T. Cole describes what this land used to be, what it can become once again, and the steps necessary to make that transition. This is more than just another account of nature's resiliency. It is a road map for enlisting the largest agricultural acreage in the country in the fight against climate change, biodiversity declines, and soil loss, all while remaining productive cattle lands. If you are looking for an upbeat view of the future, start here."

—Douglas W. Tallamy, author of *The Nature of Oaks*

RESTORING THE PITCHFORK RANCH

A. THOMAS COLE

RESTORING THE Pitchfork Ranch

How Healing a Southwest Oasis Holds Promise for Our Endangered Land

THE UNIVERSITY OF ARIZONA PRESS
TUCSON

The University of Arizona Press
www.uapress.arizona.edu

We respectfully acknowledge the University of Arizona is on the land and territories of Indigenous peoples. Today, Arizona is home to twenty-two federally recognized tribes, with Tucson being home to the O'odham and the Yaqui. Committed to diversity and inclusion, the University strives to build sustainable relationships with sovereign Native Nations and Indigenous communities through education offerings, partnerships, and community service.

ISBN-13: 978-0-8165-5280-1 (hardcover)
ISBN-13: 978-0-8165-5282-5 (ebook)

Cover design by Leigh McDonald
Cover photo by Alicia Arcidiacono, windmill from engraving by David Wait
Wood engravings by David Wait
Designed and typeset by Leigh McDonald in Corona 10/14.5 and P22 Franklin (display)

Library of Congress Cataloging-in-Publication Data
Names: Cole, A. Thomas, 1947– author.
Title: Restoring the Pitchfork Ranch : how healing a southwest oasis holds promise for our endangered land / A. Thomas Cole.
Description: Tucson : University of Arizona Press, 2024. | Includes bibliographical references and index.
Identifiers: LCCN 2023028989 (print) | LCCN 2023028990 (ebook) | ISBN 9780816552801 (hardcover) | ISBN 9780816552825 (ebook)
Subjects: LCSH: Restoration ecology—New Mexico—Pitchfork Ranch. | Restoration ecology—New Mexico—Grant County. | Grassland restoration—New Mexico—Pitchfork Ranch. | Grassland restoration—New Mexico—Grant County. | Riparian restoration—New Mexico—Pitchfork Ranch. | Riparian restoration—New Mexico—Grant County. | Pitchfork Ranch (N.M.)—History.
Classification: LCC QH105.N6 C65 2024 (print) | LCC QH105.N6 (ebook)
LC record available at https://lccn.loc.gov/2023028989
LC ebook record available at https://lccn.loc.gov/2023028990

Printed in the United States of America
♾ This paper meets the requirements of ANSI/NISO Z39.48-1992 (Permanence of Paper).

Dedicated to

Ella Jaz Kirk, Ella Meyers, and Michael Mahl—

Silver City, New Mexico, students lost in a 2014 plane crash—

to the world's young people, who are the future,

and to my little brother, Danny,

and Jake and Violet,

who didn't live to see the mess we've made of things

CONTENTS

PREFACE

THERE IS A LINK between the early cattle-ranching world and today's climate, species extinction, and soil loss and depletion crises that will become clear in this book. The Pitchfork Ranch is a platform for the book's message. It was the home and workplace for pioneer-settler Jerimiah McDonald and later for his three sons, Bartley, Taylor, and Jonnie. Taylor wrote a poem, published in the *New Mexico Stockman* shortly before his death, that captures a cowman and his family's life, forecasting the planet handed down to their children and the rest of us: hotter, dryer, damaged, and dependent on prompt solutions.

The Drouth and Me by Taylor McDonald

The feed's all gone, the creek's dried up, not much water in the well,
There's a hot wind that's a blowin' straight from the doors of hell.
I wipe the sweat from my fevered brow, I glance up at the sky,
I see them clouds a rollin', them empties goin' by.

My overalls are gettin' thin (they're the dashboard type you know)
Them fancy riden' britches I wore out long ago.

My hair's grown down the collar of a shirt with ragged sleeves,
And don't you think that Mama hasn't felt this squeeze.

She may not be the cutey that I married long ago,
But she's sure 'nough the big wheel that it takes to run this show.
Her hair's gettin' kinda gray, her face is full of seams,
But she keeps the kids in feed sack drawers and hoghide in the beans.

I came in all disgusted and swear I'm goin' to quit,
And Ma she gets so mad at me she almost has a fit.
Says, "Pa, where's your gumption, we're stuck in this here game,
We'll save some seed if we have to feed a little of Ezra's[1] grain."

There's a bright star shinin', a shinin' way up high,
Cause it's hard to see it with these empties goin' by.
Some day we'll have fat cattle, we'll have money in the bank.
We'll have the kids and you and me and we'll have the Lord to thank.

"Stay with the cow and she'll stay with you," my grandpa always said,
Now maybe times have changed some in the years since he's been dead.
But we'll stay with the cows until the last one's gone, and then if it don't rain,
We'll make up a batch of whisky of what's left of Ezra's grain.

1. Ezra was Ezra Taft Bensen, the secretary of agriculture during those drought years. He oversaw a program to assist ranchers by providing inexpensive feed for their cattle.

ACKNOWLEDGMENTS

THE LAND DISCUSSED IN this book once belonged to the First Peoples for whom this acknowledgment is faint praise. Lucinda Cole, my restoration partner and in-house editor, receives top billing for having indulged me for this decade-long effort. She made valuable comments, as well as supplying support and encouragement.

Great appreciation goes to my parents—Arnold and Muriel Cole—for the opportunity they gave me to flourish and for the way they raised us kids.

I also extend my thanks to Mary Ellen Corbett, who helped me begin this book and suggested the title. Abundant thanks to Charlie McKee, who edited and reedited every word, and to David Wait and Ellen Soles for their thoughtful support and work on the manuscript.

RESTORING THE PITCHFORK RANCH

Introduction

RUNNING FREE THROUGH THE woods in boyhood set a hook in me I've never spit out. It was the summer of 1957, I was eleven, and Buddy Holly and the Crickets had broken into song in a big way with the 45 rpm "That'll Be the Day" and follow-up "Peggy Sue." My father partnered with a neighbor and took my mother, two sisters, and me to harvest apples at Pendley's Orchard in Oak Creek Canyon, a quiet, fertile, and picturesque oasis in the highlands between Flagstaff and Sedona, Arizona. It was a special childhood summer: exploring off-limits places, hell-bent for leather, and roaming wild-like in the mountains.

My parents, sisters, and I lived in a 1930s workers' quarters; the girls slept in the kitchen, and I slept on a screen porch. Our main tasks were readying the orchard for harvest and picking apples that we packed for shipping in a run-down shed on the farm. My dad opened a fruit stand near Slide Rock. He and his partner Bill hooked up a horse to an old two-handled walking plow, furrowing a field while Mom, Bill's family, my sisters, and I planted eleven acres of the best sweet corn I've ever tasted. If we wanted more corn during supper, we'd just step outside and pick whatever we liked.

I caught my first fish, picked wild raspberries and blackberries, gathered plums, peaches, and apricots for sale at Dad's market. I smoked my first cigarette and discovered Playboy foldouts stapled to the ceiling of an abandoned farmhouse. I snuck bareback rides on a horse named Lucky, swam at nearby Slide Rock, and went where I wasn't supposed to go; I walked high above the ravines like a circus performer along a dilapidated metal flume that had carried water over deep canyons to the apple orchard for decades. After planting his first tree in 1912, Frank Pendley built a large shed, where apples were packed and readied for transport. We followed his lead. I stamped shipping boxes with the apples' name and size.

Favored with this idyllic childhood summer away from the city, school, and sports, I'd never experienced wilderness and such freedom. But at summer's end, I returned to the life of an adolescent urban dweller: baseball, rock 'n' roll, and eventually basketball, cars, and girls. After exhausting the athletic priorities of youth, I retooled: prioritized education and my future, married, and settled into career and family. I tucked away memories of that special freewheeling summer for nearly half a century as Lucinda, my wife, and I pursued careers, took part in our community, and raised our children. Places can shape us. After retiring to the Pitchfork Ranch in southwest New Mexico, I realized that the experience of my boyhood summer in Oak Creek Canyon was central to the course of my life and had played a pivotal role in the decision that brought Lucinda and me to live out our time on the ranch.

Looking back on our lives, we can identify two or three events that shaped the trajectory of our futures. I'm not talking about the touchstone events that shaped our peer group generally, those circumstances that fashioned us as a generation and occupy the collective unconscious of most of us who came of age in a certain period—like the sixties and Vietnam for Lucinda's and my generation or the depression and World War II for our parents—I'm referring to those experiences that define us as individuals. One of those events occurred when I stood quietly beside my freshman college basketball coach and another professor discussing lawyers, a conversation that interested me in a career that ultimately allowed us to retire to the ranch. Yet it's the picturesque wildlands of my idyllic youth that led us here.

Readying to leave behind our workaday worlds, Lucinda and I began our search for wilderness to purchase and restore. We soon realized that our ideal of restoring "wilderness" was not to be. True wilderness—uninhabited, arguably undisturbed, legally protected habitat—is owned by the public. It's not for sale. It turns out that the most suitable property available, similar to wilderness and in need of restoration, was ranchland. We changed our focus. The Pitchfork Ranch was the first property we visited, but we decided against it for reasons too embarrassing to tell. After a year of searching and visiting twenty other ranches, we realized our mistake and returned to the Pitchfork. We found it in escrow yet were fortunate enough to negotiate an agreement with the prospective buyer, stand in his shoes, and close the escrow for ourselves on a place to manage, restore, and preserve.

"Great strokes of luck are usually disastrous," wrote Germaine Greer in *White Beech: The Rainforest Years*, about her purchase of an abandoned 150-acre dairy farm of logged, cleared, steep, rocky, and worn-out country in southeast Queensland, Australia.[1] Like us, well past midlife, she has an intense love for Earth, and like us, her stroke of luck in finding and rehabilitating the place was anything but disastrous. Greer maintains, as we do, that there is a deep-rooted satisfaction in helping to rebuild wild nature, something special about restoring damaged land. Aldo Leopold was one of the first to recognize this and wrote about it in *A Sand County Almanac*. My goal is to capture that satisfaction while making the case that habitat restoration is essential if we are to survive the blight of global climate breakdown, species extinction, and soil loss. There has never been anything more threatening to humankind than these rapidly worsening crises. Yet each of us can respond to these emergencies by restoring our own land or signing up with a community restoration group, changing our way of life, and joining the Voice of the Streets.

The Pitchfork Ranch—in cattle production for more than a century—is now an enterprise for habitat restoration, introduction of at-risk species, carbon sequestration, research; a place for wildlife to breed, birth, and raise their young; and home to a small cattle herd and me and Lucinda. It's located fifteen miles down a county-maintained dirt road, an hour's drive south of Silver City, New Mexico. The ranch sits in the heart of a fifty-eight-square-mile watershed. Eight and three-tenths miles of the forty-eight-mile-long Burro Ciénaga watercourse, including the key portion of the surviving ciénaga, cross the ranch, north to south.

Inexperienced for our task, Lucinda and I attended workshops on restoring habitat and explored the literature for strategies to prevent the erosional processes that had been damaging and dewatering the Southwest over the last four hundred years. We learned how to reverse the losses. Our goal was to return the ciénaga and surrounding grasslands to something close to their pre-European-settlement condition. As the restoration progressed, we realized our efforts provided an opportunity to do more. Not only could we pursue restoration and provide habitat for wildlife, but the ranch offered us an opportunity to address the climate crisis, species extinction, and soil loss.

Our work has us walking in the shadows of Aldo and Estella Leopold and Wendell and Tanya Berry, repairing land and remedying its past misuse. The fundamental obstacle to be overcome is the misfortune of humankind living carelessly on Earth, disconnected from its past, detached from the natural world, and lost in a broken relationship between people and place. When Aldo Leopold and his family and friends began restoration on his and Estella's abandoned eighty-acre Wisconsin farm, and Wendell and Tanya Berry began their reclamation efforts on their washed-out farm along the western bank of the Kentucky River, they saw the hope in abused soil and an opportunity to repair a wounded place. They were unaware there was a far more pernicious process in the making, awaiting their descendants. The human-caused climate and companion crises have arrived. We're in the midst of an ongoing and escalating catastrophe. Time is critical; pause and delay have complicated efforts to address these emergencies, quickening the pace toward zero hour.

The more we learned about this land and the ranch's history, the more we came to understand the importance of *place*. We knew the literature of *placelessness* and *sense of place*, long ago having read Wallace Stegner's writing about "boomers" and "stickers," the former being the takers who leave and the latter being "those who find a place and make it a home, stay in it and try to leave it a little better than they found it."[2] We had read Scott Sanders's *Staying Put* and wanted to. We knew of the importance of finding one's place, sticking, and settling in. The significance of place—in the manner Henry Thoreau, Robert Frost, and Wallace Stegner used it in reference to a "placed" person, as well as Wendell Berry's belief that "if you don't know where you are, you don't know who you are"—was important to us.[3] Yet, living on the ranch, we came to know the intangible value of achieving intimacy with a place and working with something that matters, well beyond our earlier understandings. Our admiration for this place has continued to deepen. Places have a way of drawing you home.

Simplifying the challenges posed by Earth's crisis into a single paragraph ensures the transgression of omission, but here is a smattering of the core phenomenon conspiring to wreak havoc on the planet. Earth's temperature is higher than any time in human history. We have just endured the hottest decade on record; the Gulf and East Coast of the United States are awash in drastic weather. People are drowning in their basements, burning to death as they flee. A "heat dome" cooked British Columbia, Oregon, and Washington, killing 500 people in Canada, at least 95 in Oregon, and possibly more than a billion marine animals globally. California and Australia are ablaze, fires never more extreme or this frequent. California has suffered severe drought, then horrendous flooding—an environmental whipsaw. One-third of Pakistan was under water, affecting 33 million people. Human population growth, resource extraction, and consumption are headed toward tipping points. Earth's species are undergoing a rate of extinction from 1,000 to as much as 10,000 times higher than normal. Soil is depleted and washing away. Humankind is ensnared in an accumulation of climatological and geological changes that are modifying the basic physical processes of the planet.

We must do more than insulate ourselves from the real and present dangers of an overheated climate: rising seas, human migrations, resource depletion, the spread of infectious disease, global food shortages, and pandemics. The future must be more than isolation or the illusion of escape. Each of us must answer the questions of how to address our part in these crises and how to live on a finite planet.

Abusing land anywhere has harmful consequences for people everywhere. It's time we realize the ways the planet has turned against us—grown smaller and angrier and more erratic—changes that are human creations. University of Notre Dame professor Roy Scranton writes, "The time we've been thrown into is one of alarming and bewildering change—the breakup of the post-1945 global order, a multispecies mass extinction, and the beginning of the end of civilization as we know it. Not one of us is innocent, not one of us is safe."[4] What we do now will determine if humankind is capable of overcoming the civilization-scale threat of accelerating climate breakdown and determine "what the world will look like for thousands of decades to come."[5] We are living the reality of Christian Parenti's judgment; we are faced with "the most colossal set of events in human history: the catastrophic convergence of poverty, violence and climate change."[6] In a 2020 virtual address at Columbia University, António Guterres, United Nations secretary general, summed up our dilemma:

> Humanity is waging war on nature. This is suicidal. Nature always strikes back—and it is already doing so with growing force and fury. Biodiversity is collapsing. One million species are at risk of extinction. Ecosystems are disappearing before our eyes. . . . Human activities are at the root of our descent toward chaos. But that means human action can help to solve it. . . . Making peace with nature is the defining task of the 21st century. . . . It must be the top, top priority for everyone, everywhere.[7]

The following year, Guterres called the nearly four-thousand-page Sixth Assessment Report by the United Nations International Panel on Climate Change (IPCC)—based on fourteen thousand peer-reviewed studies—a "code red for humanity," as it makes clear the global crisis is now "inevitable, unprecedented, and irreversible."[8] The next year, he described the 2022 IPCC report as an atlas of human suffering and a damning indictment of a failed climate leadership.[9] The phrases "collective suicide" and "ecosystem meltdown" have entered the lexicon. One group of IPCC scientists initiated a petition in 2022 urging colleagues to abandon participation in this world's most important climate assessment because policy makers were paying it so little attention. People are being driven to grief and despair. Scientists have formed Scientist Rebellion, locking themselves to bank entrances. Young people are gluing themselves to streets and priceless works of art. Earth scientist Rose Abramoff was fired from her job at the Oak Ridge National Laboratory in Tennessee for the nonviolent unfurling of a banner that read "Out of the lab & into the streets" at the world's largest meeting of scientists who study Earth and space—the American Geophysical Union—and more and more scientists are joining Scientist Rebellion, an international network of more than five hundred scientists protesting globally to raise awareness of how luxury air travel and other ubiquitous conveniences contribute to the climate crisis.

Although times are bleak, climate scientist Michael E. Mann maintains the key to survival is for us to recognize

> there is still time. Don't let the "doomers" convince you it's too late to do anything about climate change. That leads us down the same path of inaction as outright denial. And the inactivists, the forces of inaction, would love nothing more than for environmental progressives to remain on the sidelines because they're convinced it's too late to do anything. It isn't too late, but we have to act, and we have to act now.[10]

Scientists, activists, and untold others concerned about these crises are calling for real reductions in carbon emissions and consumption and for real solutions, rapidly and permanently. We find ourselves at a historic crossroads. My hope is that an appreciation of these crises will deepen with this survey of the science, the details of our restoration work, and the weaving of this material into the fabric of the Pitchfork Ranch. My thesis is fivefold: (1) the overarching contention is that global warming, species extinction, and soil loss are a planetary emergency, a catastrophic threat far worse than any previous challenge faced by humankind; (2) we have failed to take these crises seriously; (3) the key solution to these crises is to change our culturally inherited, convenient way of living; (4) one of the more important responses is to pursue natural climate solutions and restore habitat; and (5) we must participate in the Voice of the Streets, instigating political agitation to transform our destructive way of living to maintain our lives sustainably.

Identifying why people write, George Orwell singled out four motives, one of which captures the primary impulse that motivated my writing this book: "desire to push the world in a certain direction, to alter other people's idea of the kind of society that they should strive after."[11] I aim to create a coherent and intellectually satisfying account of the nature, extent, and causes of these crises in the context of the Pitchfork Ranch, to detail a way to overcome the threats of climate and environmental breakdown, and to identify nature-based solutions everyone can adopt.

With the world warming, seas rising, species going extinct, soil loss, pandemics, water shortages, and ecosystem function diminishing at an alarming rate, we are facing multiple crises that are apocalyptic in scale. Scientists are warning of cascading environmental collapse. In the original sense of the ancient Greek word *krisis*—the turning point in an illness when a patient either dies or recovers—our planet Earth is the patient, yet it's her inhabitants who may not pull through. I propose the concept of survival for an expanded land ethic, a more fundamental tenet to contend with these crises. Survival requires that we share one another's fate and become a part of, rather than master of, the planet. There are answers to the tangled question: "With such complexity, what can I do now?"

CHAPTER ONE

The Pitchfork Ranch

THE PITCHFORK RANCH IS easy to pass by on the way to somewhere else, yet it's a special place with a story to tell. Since our purchase of the ranch in 2003, Lucinda and I have pursued habitat restoration born of hope. The work offers an answer to what each of us can do to assume responsibility for our part in creating the climate, species extinction, and soil crises. Undoing the human-caused damage to the planet will take a Herculean effort: prompt, bold, innovative, socially just action, and constant, disruptive, and dedicated activism. A significant slice of the challenge consists of natural climate solutions, of which habitat restoration is among the most promising: an essential strategy to combat these crises that can be embraced by everyone, wherever you live, even if you don't own property.

The history of this ranch is steeped in Western lore, archaeology, plant and animal life, and a unique hydrology—a rare source of arid-land water called a ciénaga. It's also a storied place where McDonald grandkids, many of whom spent summers here, now adults, return to visit with tales of spaceship sightings ("all five of us saw it"), cattle rustling, hidden cousins due to a long-unknown child-out-of-wedlock, adoption and her offspring, killings and more. Although humans have lived on these lands for more than ten thousand years, let me begin the story with the people of Euro-American ancestry who made this place their home. This ranch, like the Old West it reflects, is as rich in myth, conflict and tradition, as entangled in the gap between truth and story as this part of the nation itself, rooted in the history of dispossessed First Peoples, westward expansion, corporate ranching, the never-give-up, hardworking cowman, and the myth of the iconic American cowboy. The Pitchfork Ranch is a split-off portion of what had long been known as the McDonald Ranch after its founder Jeremiah McDonald; following his death, the McDonald Brothers Ranch after Jeremiah's three sons; and finally, a portion of it became the Pitchfork Ranch after the brand of Jeremiah's oldest boy, Bartley.

A year after his mother died in 1872 and a quarrel with their housekeeper over her treatment of his younger brother, fearing punishment from his father, Jeremiah McDonald ran away from his Illinois home and went to Nebraska where he found work as a horse wrangler. Several years later, he returned home, but soon he and his brother Tom left for Kansas where they tried to raise cattle but lost them in a freeze. The two boys then worked their way toward the Arizona-New Mexico Territory as wagon-train guards, arriving in Texas where they worked a short time before parting ways. Jeremiah signed on as a guard with a migrant team from Sherman, Texas, heading west for Silver City in what is now New Mexico. The guards were replaced by soldiers who met the wagon train at Fort Cummings for escort to Silver City. Out of a job, Jeremiah went to nearby Deming, New Mexico. There he was approached by a stranger who offered him work on the Cloverdale Ranch near the Mexican border, seventy miles south of today's Pitchfork Ranch.

Jeremiah McDonald joined millions of Americans in the westward migration after the 1803 Louisiana Purchase. By 1840 nearly seven million Americans—40 percent of the nation's population—lived in the Trans-Appalachian West. The United States acquired the far west region under the terms of the 1848 Treaty of Guadalupe Hidalgo, which ended the Mexican-American War, and through the 1853 Gadsden Purchase. An uptick in migration followed the end of the Civil War. Just kids, Jeremiah and his brother must have been bitten by the bug that came to be known as "Manifest Destiny," inspiring a flood of American hopefuls to head west. Half a century after Jeremiah landed in New Mexico, Franklin Delano Roosevelt looked back and recounted for a San Francisco audience what he and many others thought distinguished America from other nations. From its earliest days, "at the very worst, there was always the possibility of climbing into a covered wagon and moving west where the untilled prairies afforded a haven for men to whom the East did not provide a place."[1]

In his new job, Jeremiah wrangled horses for Tom Ketchum, who became known as "Black Jack" Ketchum—gunman, killer, bandit, and train robber—and led "the worst gang of outlaws that ever infested the Southwest."[2] He was an assassin of the worst order, killing at least half a dozen men and leading his gang in multiple train robberies. His last holdup was in Union County, New Mexico, where his brother was killed, and Ketchum was shot, captured, convicted, and hung at Clayton, New Mexico.

Still growing grass and raising cattle with her husband today on the McDonald Ranch adjacent to the Pitchfork Ranch, Linda McDonald tells this story about Grandpa Jeremiah's introduction to Ketchum and his gang:

> The boy didn't know who he was working for, but he saw a big herd of cattle come in, and there became a big gunfight going on, and he and the cook got on top of this Spanish-style roof where they were protected because the way Spanish-style houses were built. Men on both sides were shooting their guns when the cook turned to Jeremiah and asks, "Do you know who you're working for?" Jeremiah says, "No, I don't." And the cook says, "You're working for Black Jack Ketchum! If you'd just take my advice, you'd get out of here!" So, he did. He slid down the roof, grabbed a horse and got away.[3]

Lucky not to have been caught up in the Cloverdale shootout, Jeremiah landed near Cherry Creek in the Big Burro Mountains, north of today's Pitchfork Ranch. Jeremiah initially drove a stagecoach, but soon began a thirteen-year stint with the Lyons and Campbell Cattle Company. By the time he was eighteen, Jeremiah was wedded to the LC Ranch and worked for the LC until 1896.

Around the time Black Jack Ketchum and his gang were in their train-robbing prime, Jeremiah's future wife, Mitchell Ann Gordon, and her family left South Carolina for Colorado, then Mexico, and finally Red Rock, New Mexico. They settled forty miles west of where she would meet and eventually marry Jeremiah. The newlyweds worked and lived on the JO Bar Ranch near Lordsburg, south of the future McDonald Ranch.

One-armed Civil War veteran George Hornbrook also worked at the JO Bar Ranch, and despite his disability, he could still cowboy, and that's how he made his living. Jeremiah and Mitchell Ann's first child, Bartley McDonald, was born in 1899, and four siblings followed: Taylor, Katie, Jan, and Jonnie. Hornbrook took a liking to the McDonalds' eldest son, gave little Bartley a heifer, and registered the three-pronged "Pitchfork" brand in the boy's name. The name and brand stuck and remain in service today.

In 1903, while working the range, gathering and branding cattle for the JO, Jeremiah spotted an abandoned piece of land and an old adobe home with thick, port-holed walls. Taking the first step in establishing the McDonald Ranch, he paid $17.80 for the tax title to 160 acres of patented (deeded) land not far from today's Pitchfork Ranch headquarters. After relocating his family, he began acquiring the nearby failed and abandoned homesteads that eventually made up the McDonald Ranch. Reaching fifty thousand acres in size, it was one of the largest ranches in Grant County.

As a youngster, Bartley McDonald began working on John Turner's ranch between Silver City and Deming as a way to earn money to help stock cattle on his parents' place, which he and his two brothers eventually inherited. John Turner was also the Grant County sheriff, and Bartley helped him as a jailor. He worked in Silver City law enforcement and was eventually elected Grant County sheriff. Bartley served only one term, losing his office due to his role in a labor dispute that rocked Silver City and the entire Southwest in the early 1950s (discussed in Chapter Ten). While working the ranch with his brothers, Bartley continued his career as Silver City police chief, Grant County treasurer, Grant County assessor and served twelve years as a county probation officer before retiring to the ranch.

There's a tension between myth and reality in much of the history of the American West that also surfaces in looking at the past of this ranch, especially Bartley's story. It's a commonly held belief by many Americans that homesteading pioneers and the ranchers who followed were the salt of the earth, men of astounding virtue, who later scholarship maintained were the source of the American character. When twenty-five-year-old Theodore Roosevelt went west with a sizable inheritance in 1884, he fell in love with ranching and cowboys, lionizing the ranch hand as a tough, hardworking, independent, fun-loving, honest, self-confident, self-reliant, and decent human being. But recent historians contend much of the West was settled primarily by people filled with greed, corruption, violence, and racism. Discussed later in this chapter, today's historians maintain most cowboys were a sad lot: malnourished, living in miserable conditions, seasonally employed, poorly paid, scraggly, dirty men in tattered, ill-fitting clothes, solitary drifters, some of whom were the dregs of society. There were exceptions, but on balance they were a troubled lot, unlike the idealized characters who were mythologized in the Wild West shows, dime novels, black-and-white movies, and modern films like *Lonesome Dove* and *Dances with Wolves* and now *Yellowstone*.

Far removed from the life of a drifter and hardly a myth, the dominant personality of Pitchfork Ranch history is onetime Grant County sheriff Bartley McDonald. Like many characters of Western lore, the book on McDonald is mixed. A local author from the period wrote that Bartley was a *wheelhorse*. In the time of freight outfits, two of the best and most dependable horses were hitched directly to the wagon and were called *wheelers*. Indispensable, they were the ones that could be depended on to put every ounce of strength, as well as every bit of intelligence, into the task at hand and to do so every time they were called on. During this era, when someone referred to another person as a wheelhorse, they were speaking in the most complimentary terms. The term was said to "fit Bartley McDonald to a T."[4]

When he was a young man, our elderly neighbor's father-in-law saw Bartley muscle and jail two strapping roughnecks causing trouble in Silver City's Buffalo Bar. Bartley grabbed both men by the nape of the neck, almost lifting them off the

FIGURE 1 The McDonald Brothers: Bartley, Taylor, and Jonnie in 1947. (Courtesy of C. E. Hellbusch, Phoenix, Arizona; photo by Mickey Lemon)

ground, hauling them to jail up Bullard Street. In keeping with the "wheelhorse" remark, his daughter remembers him as a man with a presence to be reckoned with, who believed in self-discipline. She maintains the youth he dealt with, who were protesting life in the sixties, came back years later and thanked him. The word was, when a troubled kid left his office, he never left without money in his pocket.[5]

There's another, less-varnished, view of Bartley McDonald. His niece told me he was too hard on his son, Pete. Recently deceased Bobby Sellers, a longtime Silver City electrician who helped us restore the Pitchfork Ranch headquarters, also understood Bartley's legacy differently. Bobby grew up in the Silver City neighborhood where Bartley lived, and there was never any question that he was the law. Kids were terrified of him, not because he was mean—Sellers says he never saw that in him—but because he was big and tough, and he was the sheriff. "Bartley McDonald was both feared and respected."[6] Bobby also recounted long-ago lore that Bartley killed two blameless "wetbacks" who worked for him and claimed to be owed thirty-seven dollars. As Sellers understood the tale, they confronted Bartley in the small town of White Signal, where we pick up our mail, fifteen miles up the road from the ranch. They ended up dead. Only Bartley had a gun.

The *Silver City Daily Press* accounts from the period disclose a criminal complaint for murder being filed against McDonald, although the charges were dismissed before the preliminary hearing on the grounds of self-defense. Dana Forntenberry, who helps us with our electrical and water systems, recounted a run-in his parents had with Bartley when he was a young boy. When I mentioned Bartley, I could see the veins in his neck tighten as he became serious, speaking in a way suggesting he had felt this way for a lifetime, "The son-of-a-bitch died of a mean heart."[7]

Bartley was twenty-seven years old when he met his future wife, May Beth, who was only fifteen years old at the time. Granddaughter of a polygamous Mormon family from the enclave near Casas Grandes, Mexico, she and her family were forced out of Mexico when Poncho Villa ordered all White settlers to leave within the hour. When Bartley asked May Beth's father for her hand, he answered, "OK but you'll be waiting until she's seventeen. Well enough, he did."[8]

George Snyder arrived in the area in 1885 and formally homesteaded some of the land that became part of the McDonald Ranch. By the time of his death, the Snyder Ranch had grown to about twelve thousand acres, six thousand of which were patented land and the balance federal government and state of New Mexico leased land. When Snyder died, Bartley, Taylor, and Jonnie McDonald and their cousin J. L. McCauley purchased the Snyder Ranch and established the McDonald Brothers' Ranch that eventually incorporated their father's place when he died.

Working the ranch was hard labor and a lonesome, painstaking grind. Jonnie, Jeremiah's youngest son, remembers early ranch life as a time dominated by long stretches of isolation. Jeremiah started with near nothing, worked the roundups on other ranches, and was often gone for months at a time. Jonnie told a high school journalism student that his mother would not see her husband or anyone but her own kids for "as high as three months at a time . . . so she was kind of a rugged character also." Back then, there were few visitors: "they wasn't anyone to visit, no one to see. All us kids was scared to death of wimin. There wasn't even ever a wimin showed up. Why if ever one did, we would run. We wasn't ever afraid of men 'cause they was quite a few men came around, line rider and so forth. But we didn't ever see many wimin."[9] Jonnie McDonald and his brothers acquired more pieces of the ranch and purchased the final two 480-acre parcels of land in 1961, completing what by then was the much larger McDonald Brothers' Ranch that today has been split into three cattle operations: the McDonald, Thorn, and Pitchfork ranches.

Establishing the Pitchfork Ranch, like many of the ranches in the history of the Southwest, involved acquiring abandoned ranches when earlier settlers were "starved out." Most of the predecessors to the McDonald Ranch are lost to history, as they gave up and joined other disillusioned pioneers who were drawn west by the mythic 160-acre quarter section of land. The homestead laws, enacted by legislators whose livelihoods were made with land blessed with plentiful rainfall east of the hundredth meridian, failed to establish a settlement policy that allowed settlers to acquire enough land to support a family in the arid country farther west, land with little and unpredictable rain. Unable to make a living on small spreads, many settlers were forced to cheat in order to acquire more land and often still had to sell their homesteads to neighboring ranchers lucky enough to have a source of water. Or they simply walked away and deserted their dreams, abandoning their homesteads owing property taxes. It was nearly impossible to avoid the failure assured by the combination of the tiny quarter section and aridity. The Burro Ciénaga watercourse allowed the McDonalds to have six wells all less than one hundred feet deep, assuring their survival.

When Wallace Stegner wrote that the West was a place where optimism consistently outruns resources, that inherited wetland habits from the East were in conflict with the dry lands of the West, and that aridity called the tune, he summarized the land's delicacy with this rephrasing of John Wesley Powell's forewarning: "[The] compelling fact is that the basic resources of water and soil, which can be mismanaged elsewhere without necessarily drastic consequences, cannot be mismanaged in the West without consequences that are immediate and catastrophic, and that reach a long way."[10]

The difference between the annual forty inches of rainfall in areas east of hundredth meridian and twelve inches of rain in the American Southwest is a knottier imbalance than just a four-to-one ratio. Aridity causes snowpack to melt faster, less forest cover leads to faster melting snow, and evaporation rates are higher. So yes, aridity calls the tune, and vulnerability is what you get. As we'll see, the difference in rainfall west and east of the hundredth meridian defined much of John Wesley Powell's carrier, dominated a considerable portion of Stegner's most influential writing and provides us with a critical warning posed by the ongoing and rapidly expanding threats of the environmental crisis.

In 1958 fifty-six-year-old middle brother Taylor died of a brain aneurysm sustained in a horse wreck, leaving his wife, Florence, as co-owner with his brothers. Taylor and Florence's children told me they loved their life on the ranch and assumed they would always have the ranch to come home to and visit, even though they knew it was not big enough for the entire family to make a living. The changes that followed their father's death pulled the rug from beneath their branch of the family. Taylor's youngest daughter wrote late in life, "Thus ended an era that was cut short by an untimely death with work left undone and dreams left unfulfilled."[11]

After Taylor died, Bartley and Jonnie purchased the Thorn Ranch south of the McDonald Brother's Ranch but did so financially separate from the three-way ownership of the three brothers (now Florence, Jonnie, and Bartley), at which point the ranch reached its peak size of fifty thousand acres. Land use and allocation of profit between the two operations became a difficult egg to unscramble, so Florence sold out to Bartley and Jonnie who eventually sold the Thorn Ranch, intending to pay Florence over a period of years. Whether or not they did is unclear. Taylor's daughters told me that Jonnie was kind beyond measure, Bartley independent and difficult, and Taylor the glue that kept things running—not always smoothly, but running. The daughters believe the final split may have been inevitable after their father's death. When the remaining two-thirds ownership was finally divided as it remains today, one account is that it was divided so Jonnie's grandchildren would be closer to school; another version says Jonnie abused a horse, causing Bartley to anger and leading to the final split of the ranch in 1976. The reality of family ranching is that most ranches are large enough to provide for one family, but not the next generation of multiple families, leaving fertile ground for tension, conflict, and disappointment.

When Jonnie and his family moved north, he took the McDonald Ranch name, still in use today by his son, Jerry, and wife Linda. Bartley retained the Pitchfork brand given to him as a youngster by the one-armed Civil War veteran. By the time Bartley reached the age of seventy-six, his daughter said her dad felt he was running out of time.[12] After the original McDonald Ranch was divided as it remains today, he retired, leaving public life. Within a year, Bartley was dead. No one knows for certain the details of how he died; stories vary. Bartley had gone to the pasture near Soldier's Farewell Hill to retrieve his grandson's horse, Park's Pacer. Some believe the stallion charged Bartley's mare, throwing him off the horse and causing fatal head injuries; others speculate that his favorite horse, Paymaster spooked, afraid of snakes and jackrabbits.

Bartley's daughter Patsy recalls her mother, May Beth, went to town to buy provisions for the next day's roundup. She arrived at home late, Bartley had not returned. She knew he had gone to Soldier's Farewell Hill to retrieve Park's Pacer, so she and other family members drove there to check on him. Bartley's truck and trailer were in the cow lot, but there was no sign of him. They spread out and searched without success. They collected firewood and built a huge fire, thinking Bartley might see it in the dark. His son, Pete, arrived at the ranch after nightfall and searched until the sun came up, also to no avail.

"At daylight, they found him. His glasses were still on his face, and his rope was in his hand."[13] Linda McDonald recalls: "He evidently was alive when he hit the ground because he was lying the way he always used to take a nap when they were working the fence lines. He was lying with his legs crossed at the ankles, and he had a rock under his head for a pillow. That was just something he did. But he was dead."[14] May Beth sat with Bartley in the bed of a pickup truck in the pouring rain on the trip to the headquarters where they waited for the coroner.

While differing accounts of Bartley's fall and by whom he was found persist, his death led to the Pitchfork Ranch forever leaving the McDonald fold. After he died, May Beth, their son, Pete, and other family members continued working the ranch, but when Pete—only forty-five years old—was taken by an aneurysm caused, like his dad, from a horse-riding accident, May Beth eventually let the ranch go. She sold it to a neighboring ranch family who owned it for twelve years and then resold it to an Arizona farmer in 1998. He married a Texas woman who wouldn't abide such a modest home, so he sold it to Lucinda and me in 2003 and moved with his bride to Texas.

This was rough country. Two of three McDonald brothers and one son were killed in horse wrecks, and Jeremiah died of pneumonia after helping a neighbor with his cattle in a winter storm. Two of the twelve original settlers whose lands were acquired to make up today's Pitchfork Ranch portion of the original McDonald Ranch were murdered. There is no longer any question that the myth of the iconic cowboy was an exaggeration and the notion the cowboy is the source of the American character is part of the myth. But that is not to suggest life on the range was anything but hard labor and perseverance. In the McDonald family alone, in addition to the deaths of Jeremiah and two of his three sons, grandson Jimmy McDonald died in 1957 from lung poisoning as a result of "cow dipping," and grandson Todd died by suicide. It's said that Todd never got over the guilt from having been told as a twelve-year-old that his horse killed his grandpa. Many of our ranching neighbors are busted up and broken from riding, roping, and the other demanding chores of raising cattle.

Bartley died on his favorite place on the ranch, Soldier's Farewell Hill. At 6,173 feet, it's several feet too short to qualify as a mountain. A well-known landmark, the hill has three legends explaining how it earned its sentimental name. The most dramatic version claims soldiers manning the signal station—a heliograph station that sent messages via Morse code with mirrors by day and flares by night—were trapped by Apaches. Tormented by thirst, they signaled "farewell" to fellow soldiers and headed down to fight the Indians. They perished. Another account states a soldier was despondent over a Dear John letter and took his own life. The most widely accepted explanation says that soldiers, escorting wagon trains in route to California, were ordered to go no farther than the hill where they bid the travelers "farewell."[15] In keeping with the folklore of departure, Bartley bid "farewell" from the place on the ranch that meant the most to him.

Two other people who passed through these lands warrant mention: John Russell Bartlett, historically noteworthy, and Inez Gonzales, more or less lost to history but whose story is as curious as anyone's connected to the Pitchfork Ranch.

John Russell Bartlett's colorful and varied work life included merchant, book dealer, bank cashier, author, publisher, librarian, Rhode Island secretary of state, and U.S. boundary commissioner after the Mexican-American War. The part of Bartlett's career of interest here is his truncated stint as boundary commissioner from 1850 to 1853. The two-thousand-mile-long border between Mexico and the United States needed to be redrawn after the Mexican-American War of 1846. The postwar task was to survey the new boundary, an assignment given to the United States-Mexico Boundary Survey.

It came to light during the survey that deficiencies in the Treaty of Guadalupe Hidalgo that ended the war incorporated errors in J. Disturnell's 1847 map—inaccuracies that led to a major flare-up and threatened resumption of armed conflict. At issue was an approximate 30,000-square-mile strip of arid land that included property making up today's Pitchfork Ranch. The boundary commission found that Disturnell's map located what is now El Paso, Texas, 34 miles north and 130 miles east of its actual geographical coordinates, thereby misplacing the southern boundary of New Mexico by miles to the north.[16]

Pedro Garcia Condé and his crew undertook the same task for Mexico as Bartlett did for the United States. The two had gotten along well, and Condé accepted a compromise proposed by Bartlett that split the difference over the disputed land. The treaty empowered the commissioners to enter binding agreements on such matters so long as the head surveyors of both countries signed off. Ratification by their respective governments was not required if the surveyors agreed. Andrew Belcher Gray, the official American surveyor, refused to sign off on the agreement, believing the compromise was too far north, depriving the United States of a rail line to what many thought was "the great gateway to the Pacific Ocean."[17] Political skirmishes ensued. Bartlett continued his work, but when Congress officially refused to honor the agreement, his good name was tarnished, and he had no choice but to resign. After he stood down, the dispute was resolved by moving the line south with the United States paying Mexico $10 million under the terms of the 1853 Gadsden Purchase, rescuing the land and securing it within the border of the United States. Had it not been for the agreement, the Pitchfork Ranch land would have remained in Mexico.

Bartlett describes the rescue of Inez Gonzales in his *Personal Narrative*, where he said the events would forever be remembered by members of the boundary commission. "It was such as to awaken the finest sympathies of our nature; and by its happy result afforded a full recompense for the trials and hardships attending our sojourn in this inhospitable wilderness."[18] Bartlett still headed the boundary commission at the time of Inez's rescue at Santa Rita, thirty miles north of the future Pitchfork Ranch. She was the captive of a group of New Mexicans who arrived at Bartlett's camp seeking provisions.

Inez had departed her home in Santa Cruz, Sonora, Mexico, with her uncle and others of her family, protected by a military escort of ten soldiers under the command of a commissioned officer. They were ambushed by Piñal Indians. Her uncle and eight soldiers were killed. Inez, two other girls, and a boy were captured. In the hands of her captors, Inez suffered greatly, doing hard labor for seven months before she was sold to a group of traders led by a White man named Peter Blacklaws of Santa Fe. Bartlett's narrative avers that the band was taking Inez to some part of New Mexico with plans to sell her in whatever way to make the most money, likely prostitution or slavery. The terms of the Treaty of Guadalupe Hidalgo were explicit that purchases of the kind at issue were prohibited "*under any pretext whatsoever*."[19] Bartlett saw it as his duty to extend the protection of the laws of the United States to her and to see that she was cared for until returned safely to her parents.

Taking charge of Inez, in late August 1851, Bartlett and the commission departed their camp at the Santa Rita copper mines and headed westward for the San Pedro River in what is now Arizona. After camping for the night of August 28 at Ojo de Vaca—Cow Springs, located on the AT Cross Ranch to the northeast of the Pitchfork Ranch—they struck across the plain, due west, to pass a spur of the Burro Mountains where they approached land that would become the Pitchfork Ranch. The group traveled twelve miles to what later became known as Soldier's Farewell Hill, where they met a Mexican lancer who told them that by turning up a *cañon* or defile to the north, they would find an excellent spring, the last water for another forty miles.[20] They next entered a narrow and picturesque passage thickly wooded with scrub oaks for about five miles where the land opened to a beautiful grassy meadow about three hundred yards wide with many springs. Here they camped.

In the five-mile sojourn up the canyon, they passed the location that would become the Pitchfork Ranch headquarters, the site of the windmill shown on the first page of the Introduction. The weather was extremely warm as they traveled north, causing their young charge to suffer from exposure to the sun. In an effort to comfort Inez, upon arriving at what was then known as Ojo de Gavilan or Hawk Spring, Bartlett renamed the spring after her: Ojo de Inez. The spring is known today as Ciénaga Spring, located on the upper, northwest edge of the ranch, although the designation of Ojo de Inez can be found on older maps of the region.

Inez traveled with her rescuers for three months. She was the only girl traveling with the commission, which romanticized her as a proper Victorian lady (fig. 2). She was sketched mounted sidesaddle on a mule, wearing a mantilla on her head and an ankle-length dress. This mythologized scene occurred along the Sonoita Creek near Patagonia, Arizona, as the boundary commission made its way west to California.

When the commission arrived in San Diego, Bartlett's gesture of good will toward the young Mexican girl was publicly recognized. Commission artist H.C. Pratt painted a portrait of Inez and a First Peoples chief. It was reported that Pratt's work was to be placed in the Indian Gallery of the Patent Office in Washington City.[21] The press mentioned a beautiful portrait, executed by Pratt, of Inez Gonzales, a handsome Mexican captive girl, that was to accompany the other paintings to the capital, although the portrait of Inez has been lost.

Later that September, Bartlett, John C. Cremony, and other commission members traveled to Santa Cruz, Sonora, Mexico, where Inez was returned to her mother. She was welcomed

FIGURE 2 John Russell Bartlett, "Fording a Stream, Pack mules Sink in quick sand," September 13, 1851, 10" × 12 ¾", Pencil and Sepia wash. (Courtesy of the John Carter Brown Library, Brown University, Providence, Rhode Island)

with repeated embraces, amid tears, prayers, thanksgiving, and joyous cries. Cremony wrote, "Mr. Bartlett conceded to me the privilege of placing Inez into the longing arms of her mother . . . one of the most affecting scenes conceivable."[22] Her parents never dreamed they would again see their child, thought lost to them forever. These events are now fictionalized in the historic novel *J.R. Bartlett and the Captive Girl* by Nancy Valentine.

The 1870s and 1880s were periods of corruption and fraud in the Southwest on a grand scale. "It was a 'great barbecue'" is how historian Vernon Parrington described the post–Civil War seizure of the West in 1927, with the largest portions going to the most powerful corporations and conglomerates. "It was a splendid feast."[23] The phrase "Free Soilers" entered the lexicon: members of a mass, radical, democratic party advocating for free distribution of land from the public domain to homesteaders.

In *This Land: How Cowboys, Capitalism, and Corruption Are Ruining the American West*, Christopher Ketcham writes that "the corruption in the West was such that by 1885 *The New York Times* denounced the 'land pirates' whose 'fraud and force' had excluded the citizen settler from 'enormous areas of public domain' and 'robb[ed] him of the heritage to which he was entitled.'"[24] Historian Lynn Jacobs maintains the ruling class of the West was made up of wealthy financiers and influential investors who put up the capital to finance huge grazing fiefdoms that soon dominated most of the West.[25] Most private range land in the western states was originally obtained by exploitation of the ill-conceived Homestead Act. The cattle barons were not cowboys. In essence, the government gave or sold the nation's public domain to speculators and corporations rather than to grangers and nesters. By 1923 more than two hundred million acres intended for settlers was instead transferred to dummy agents of speculators. A sobering 95 percent of the titles to land under the Desert Land Act of 1877 were obtained fraudulently and ended in the ownership of corporations.[26] Acquisitive, opportunistic entrepreneurs were the rich and powerful bankers, lawyers, politicians, publishing magnates, mining tycoons, timber barons, railroad kings, and industrialists. They were primarily absentee owners who lived in the East or Europe and only occasionally visited the land.[27] In the best book I've read about the cowboy's world, *The Real American Cowboy*, Jack Watson writes: "Most cowboys on ranches worked for very rich capitalists. In the seventies they typically worked for a partnership of a Western owner-manager and an Eastern financier or two."[28]

The settlement laws were abused by the skulduggery of developers and speculators, unscrupulous corporate interests who developed a number of strategies to overcome what little there was of government oversight. Corporate cattle companies flourished. While settlers found themselves subject to the massive wrongs of corporate enterprise, many of them also exploited the government's misguided quarter-section settlement template. Land was purchased in the names of relatives, employees, friends, and fictional people. The idea was simple enough: by having cowhands homestead a quarter section and then transfer it to the employer, the rancher could assemble a productive cattle operation. There was also the ploy of figuring land acquisition in a checkerboard fashion—interspersing public lands in the midst of deeded property—resulting in lowering the utility and value of public lands, thereby enhancing leasing prospects for ranchers.

In the early 1890s, another one-armed Civil War veteran and director of the United States Geological Survey, General John Wesley Powell—"the last of the nation's great continental explorers and the first of a new breed of public servant: part scientist, part social reformer, part institution builder"—proposed to end the free-for-all settlement of the West, hoping to prevent water monopolies and stop the rash of failures of "the tradition-bound, hopeful, and doomed homesteaders."[29] Instead, Powell proposed government supervision of land use, guiding development that would serve the common good rather than the convenience of the well-heeled and others who profited from a period of rampant greed by speculators and railroad corporations. Powell put forward a policy designed to benefit the "greatest good for the greatest number for the longest time," in Gifford Pinchot's words.[30]

Powell had been urging change in the patterns of western settlement for more than a decade. In 1878 he published *Report on the Lands of the Arid Region of the United States*, intended to revolutionize "the system of land surveys, land policy, land tenure, and farming methods in the West, and a denial of almost every cherished fantasy and myth associated with the Westward migration and the American dream of the Garden of the World."[31] Knowing families in the semiarid West could not make a living on only 160 acres, Powell recommended 80 acres as the appropriate size homestead for irrigated farms and 2,560 acres for pasture farms, sixteen times the normal quarter section.

His reasoning was that land west of the hundredth meridian received less than twenty inches of annual rainfall, and twenty inches was the minimum for unaided agriculture. Wallace Stegner makes this boundary explicit: "The inflexible fact of aridity lay like a fence along the 100th meridian."[32]

Powell urged a shift from free land to oversight legislation and put the brakes on railroads and trusts monopolizing ownership of the new lands. His goal was clear: Powell wanted to usher in the development of agriculture in a way that would avoid the hardship that followed ineptly designed settlement policies and the failures they spawned. Powell argued, "it would be almost a criminal act, to go on as we are doing now, and allow thousands and hundreds of thousands of people to establish homes where they cannot maintain themselves."[33] But within a year of the temporary suspension of western settlement that was intended to allow surveying and planning, western politicians successfully attacked Powell, mislabeling his surveys as a scheme of geology and socialism, resuming the plunder of the West, and assuring the eventual mass failure of homesteads.

Powell was the premier organizer and champion of the first phase of government-sponsored science and is considered father of the United States Geological Survey and founder of the Bureau of American Ethnology. For many years, his thinking remained out of favor with the prevailing views of both those profiting from settling the West and the settlers themselves. Half a century after Powell's death, Wallace Stegner's biography *Beyond the Hundredth Meridian: John Wesley Powell and the Second Opening of the West* lifted Powell from obscurity, providing a thorough account of Powell's world and resurrected his reputation. History now acknowledges Powell and his ideas as prophetic. Stegner details the once-forgotten account of Powell and his science-based judgment in *Beyond the Hundredth Meridian*, arguably the most important book written about the American West. This classic exposé chronicles the exploits of Powell's Colorado River expeditions and the logic of his enlightened views for settling the West. Former Arizona governor and secretary of the interior Bruce Babbitt's praise captures Stegner's insights, their importance in understanding the West's aridity, and science's warning in the face of climate breakdown: "When I first read *Beyond the Hundredth Meridian*, shortly after it was published in 1954, it was as though someone had thrown a rock through the window. Stegner showed us the limitations of aridity and the need for human institutions to respond in a cooperative way. He provided me in that moment with a way of thinking about the American West, the importance of finding true partnership between human beings and the land."[34]

Contrary to folklore, the legendary conqueror of the fabled western frontier enjoyed this new country for only a short time. The cowboy was soon forced to look elsewhere for work when the open range was fenced in the late 1880s and the cattle business became corporate, investor driven, and less labor dependent. Most cowhands hoped to establish their own spreads, the cattle ranches of their dreams. But the large, incorporated cattle companies did to them what unregulated, corporate capitalism does to its competition today: through myriad tactics and schemes that exploited ill-thought-out government policies, they squeezed the small operations until they were run out of the independent homestead-ranching business. The brutal details are legendary, with behavior that was typically violent and successful—crimes and killing, and lots of it. It was near impossible for the small outfit to compete and increase its holdings to a profitable size for raising cattle. In Texas,

> [county systems] of registration of brands, recording of sales, and inspection of herds protected the property of cattle companies. The cattle of a small family rancher would often be legally appropriated or the family rancher arrested on trumped-up warrants by the company's or the cattle association's control of county government. Northern companies had to rely on hired assassins to eliminate nester ranchers and to

wait to seize nester cattle "legally" until their associations controlled the commissions of the state or territory.[35]

In the 1880s the Pitchfork Ranch lands were part of the enormous Gray Ranch, headquartered in the New Mexico Bootheel, and stretched from south of the Mexican-American border to almost as far north as Socorro, New Mexico. Roland Speed—expelled in 1884 from the Southwest Stockmen's Association for conduct unbecoming to a member and branded in the local press as little more than a common thief—grazed some of the first homestead cattle on land that became part of the McDonald Ranch. He put four hundred head of Texas stock cattle on the range near Soldier's Farewell Hill. Eleven more settlers followed between 1892 and 1946, and their mostly failed homesteads make up today's Pitchfork Ranch. This fifty-four-year sequence of settlers' startups and failures makes plain the hardship that Powell's proposed policies strove to avoid. Starting with the first quarter section purchased by Jeremiah McDonald for taxes, homesteaders' failures allowed the McDonalds to assemble enough land to profitably grow grass and raise cattle.

The base of an adobe chimney, foundation footings, and rockfall from a livestock corral of the initial homestead still rest on an overlook near the ciénaga, where surface water has flowed for eons. According to family lore, Jeremiah spent the night in that abandoned house—essentially gone now but still standing at the outset of the twentieth century—when he arrived in the lower Burros. About a mile south of this first homestead, walls and interior fixtures (wood boxes for cupboards, hinges, and roofing) of Ray Gunn's stone and tree-trunk dugout home still remain along Gunn Canyon, one of three major canyons that drain into the Pitchfork Ranch's most important feature, its ciénaga reach of the Burro Ciénaga watercourse. Attending a Christmas gathering at the nearby community center, an elderly neighbor told me she and her husband once leased and ran cattle on the Pitchfork. During roundup, they hired Gunn as a day laborer. She mentioned her Mexican ranch manager had had trouble with Gunn and told her, "Yo trabajo con el pistola no mas" (I'm not working with the pistol anymore).

On land that became part of this ranch, Ernest (Bart) Irwin, father of Claude Irwin, began "proving up" his homestead in the early 1920s. In 1924 he began building the house that eventually became the ranch headquarters where Lucinda and I now live. About a year into the work on his home, Bart Irwin was shot and killed by former Grant County commissioner Pierce Rice near the Rice homestead—land that is now part of the Pitchfork Ranch, two miles south of the headquarters. After his father's death, Claude completed the headquarters (fig. 3). The McDonalds enlarged the house and surrounds with an indoor bathroom, bunkhouse, tool room, and storage area. With only minor additions, we have restored the premises in keeping with the era in which it long served as the ranch headquarters.

FIGURE 3 The ranch house Bart Irwin was building when he was killed. Photographer and subjects are unknown. Courtesy of Bartley McDonald's daughter, Patsy Adams.

The Irwin killing is typical of the bloodlettings common in the post–Civil War era and a way of life that tested the mettle of Southwest pioneers and their children. Bart Irwin and his family had been living at today's Thorn Ranch, six miles south of his new home. He traveled past the Rice place to and from the work site. Rice and Irwin had children about the same age: Rice a daughter, Irene; Irwin a son, Claude. Irene was said to be "slow" due to a fever as an infant. She and Claude began a relationship that led to her pregnancy. Claude wanted to marry the would-be mother and give the child a name, but—as the tale is told—because Irene was "slow," Bart Irwin refused. Bart continued traveling past the Rice place despite warnings he was risking a confrontation. He chose not to use an alternative route and ended up dead. After the shooting, Rice rode to Al O'Brian's headquarters and told O'Brian he had shot Irwin. They returned to the scene, loaded Irwin's body into O'Brian's pickup, and drove to Silver City, where Rice turned himself in. He pled self-defense, but a jury convicted him of second-degree murder, and he was sentenced to the state penitentiary. To preserve his daughter's honor, local lore avers Rice never disclosed at trial that his daughter was with child, fathered by the decedent's son. News accounts dispute this myth of fatherly honor. Legend also declares that sheriff's deputies found a woman's footprints where Irwin was shot, leading to speculation that Anne Lee Rice was the shooter and that her husband took the fall.

Three years later, a group of neighboring ranchers traveled to Santa Fe, met with New Mexico Governor Richard C. Dillon, and told him the purported missing piece of the story about Rice, Irwin, and their children. The governor agreed to pardon Rice if he left New Mexico. He was released, and the Rice family moved to the state of Washington, where his employer and owners of the famed Diamond A Ranch (formerly the Gray Ranch) had another ranch. Although no one suggested that Rice was within his rights to kill Irwin, his actions were considered by locals as forgivable, demonstrating Old West notions of right and wrong. Nothing was known about what happened to Irene Rice or her child. Several years ago, a group of people from Deming asked if they could spread their mother's ashes on Soldier's Farewell Hill. We agreed. When I asked why their mother wanted her ashes scattered here, they said they didn't know. The answer didn't ring true and prompted the kind of speculation that creates myths: Could it be that Irene wanted to be laid to rest near where she had chanced upon and then lost first love?

Not so. In the summer of 2021, Richard Sanders, the great-great-grandson of Pierce Rice, phoned us, asking to visit the Pitchfork Ranch. He wanted to look around the homestead where Rice shot and killed his neighbor. We agreed. Irene was the great-aunt of our visitor. He recalled meeting her when he was eight years old. He recounted his family lore. Unclear to me: Was the disabled child exploited? Is this how winners rewrite history or how families tweak a shameful past? Richard's understanding of the killing was that the neighbor boy had raped Rice's daughter. Rice went out looking for the boy, happened on his father, and killed Irwin in a duel. Searching the *Silver City Press* for accounts of the shooting and trial, Richard learned Irwin was killed with a rifle, hardly a dueling gun. Richard hopes to trace down an adoption record for the child of Irene and Claude. The child's age would be consistent with the age of the woman whose ashes were recently scattered on Soldier's Farewell Hill, lending support for a legend that Irene's daughter asked to be buried near where she was conceived.

The Irwin land and home were eventually acquired by the McDonalds, and the headquarters was enlarged and a bunkhouse added as the McDonald family grew and had extra funds from a decent calf crop or a good cattle market. We've added a study to the bunkhouse and a studio near the headquarters. Nearly century-old plumbing made an outhouse necessary. Preserving the integrity of this historic period, we finished recent repairs and construction using similar materials, design, and skill level found in the original structures.

FIGURE 4 The Pitchfork Ranch headquarters today. Photograph by Lucinda Cole, 2006.

There were more killings among the initial owners of property that became part of this ranch. John and Sadie Patterson were owners with attorney Idus L. Fiedler of the Western Belle Gold Mine, located at Gold Hill near Hornbrook Mountain. The mine is fifteen miles northwest of property they were homesteading; a quarter section that is now part of the Pitchfork Ranch. In 1892 fifty-nine-year-old John Patterson and his Silver City partner were walking past the local Gold Gulch Saloon when they were accosted by one of Patterson's employees, demanding whiskey. Patterson purchased a bottle of whiskey and gave it to the miner, only to be met with his insistence he purchase a second bottle. He refused and walked on to the Patterson home, unaware they were being followed. During dinner, noise coming from the barn caused concern for Fiedler's horse. Patterson, a longtime Grant County resident and said to be a quiet and peaceful man, didn't own a gun, but Fiedler insisted that Patterson take his Colt .45 with him to investigate. Patterson approached the barn and was shot. He fired back and instantly killed his assailant. Patterson died the following day.

The ubiquity of the Colt six-shooter coincided with the hardening of soldiers by the Civil War. The combination produced an aggressive and dangerous breed of men. Their thinking infected the region, as young men unable to break their habit of killing during the war helped to increase the 1880 New Mexico Territory murder rate to forty-seven times higher than the national average. Whiskey was part of the problem, and it and the .45-caliber bullet have a curious link in the Old West. A .45-caliber bullet for a six-gun cost twelve cents, the same as a shot of whiskey. When a cowboy was short of cash, he could always barter with the saloon keeper, using a bullet in exchange for a shot of whiskey. This interplay between the bullet, the shot, and whiskey is clever enough, but adding death to the anecdote surely held little amusement for Sadie Patterson and others widowed by killings in the Southwest.

Another slice of Western history provides a scholarly backdrop for this account of Pitchfork Ranch. This sliver of yesteryear may have had little to do with how the ranch was settled, worked, and expanded, but it fostered the myth of the cowboy that shaped the world in which the McDonalds lived and likely affected their sense of who they were. In 1893 a young scholar by the name of Fredrick Jackson Turner made a presentation at the annual meeting of the American Historical Association that soon made him famous. In "The Significance of the Frontier in American History" Turner hypothesized that in the process of conquering and gradually moving back the western frontier, early pioneers created the character of a nation: "Both democratic institutions and the American character have been largely shaped by the experiences of successive frontiers, with their repeated dream of betterment, their repeated acceptance of primitive hardships, their repeated hope and strenuousness and buoyancy, and their repeated fulfillment as smiling and productive commonwealths of agrarian democrats."[36]

Turner's "frontier thesis" and his idea concerning the "closing of the frontier" coincided with the 1890 Census Bureau declaration that the West was fully settled. His audience was indifferent and next day, the press covering the conference didn't mention it. Yet Turner's thesis was soon embraced by industry, academics, politicians, just about everyone, and soon became the most influential essay in the study of the West's history, embracing, as it did, a novel and compelling explanation for Americans' view that they were exceptional—a belief that has persisted in the form of American exceptionalism. Turner eventually lost faith in his so-called frontier thesis, but it took hold to become a core belief in how Americans thought of themselves.

A scattering of historians long argued that Turner's thesis omitted critical aspects of the story of the American Southwest, yet his ideas dominated scholarship long past his death in 1932. His thesis argued that understanding America required one know the history of the West and the character of those who settled it. Turner established the central idea that helped define the American myth of the cowboy, the pioneering experience of those migrant Americans who answered the clarion call of "Manifest Destiny," and the Anglo-centric notion that the Christian God willed American expansion westward to the Pacific Ocean. Although newspaperman John O'Sullivan would not coin this now-famous phrase—Manifest Destiny—until 1845, Alexis de Tocqueville sensed the heaven-sourced summons when he visited America in 1831: "In this gradual and continuous advance of the European race toward the Rocky Mountains, there is something providential; it's like a flood of men ceaselessly swelling, drawn on . . . by the hand of God."[37]

The narrative of American settlers' populating the frontier and the mythology of the West has long served as the hallmark of individual achievement and triumph in a land of opportunity without limits. Although this fabrication has been soundly debunked in a scholarly sense, an abundance of reputable scholarship backed up the saga of the cowboy's role in

America's providence-based destiny to claim the West, and much of it persists today.

This tale of the West that most Americans grew up with was the legend of the lone cowboy as heroic pioneer, busting open a frontier filled with heathens and struggling to claim this wild, "savage" place for his own. This was a mythical frontier showcased in the dime novels by Louis L'Amour, the succession of black-and-white movies of the 1950s and 1960s, John Wayne movies, and later blockbusters. The myth of the independent, self-reliant cowboy has no equal in American iconography. Historian Richard Hofstadter wrote: "One might question one or another of the types, but no one would deny that the cowboy, in his changing forms, has entered more deeply into the consciousness of more Americans of different generations and backgrounds than any other popular types."[38]

Turner was succeeded by historians who maintained the frontier was America's creation myth. In these wide-open spaces, a new, freer breed of man was born, unbridled by pretense, caution, or tradition. The cowboy has been long understood to be hardworking, independent, and fearless. He was the tireless, self-reliant, and optimistic everyman in the new land with its promise of ever more freewheeling prosperity. I've long forgotten where I happened on the notion that there are no lies as powerful as myths and no truths more fragile than those no one wants to hear, but it captures America's paragon of the cowboy.

Suggesting Turner was a gentle man with little capacity to see the shameful side of the rush westward, Hofstadter wrote that Turner ignored the "riotous land speculation, vigilantism, the ruthless despoiling of the continent, the arrogance of American expansionism, the pathetic tale of the Indians, anti-Mexican and anti-Chinese nativism, the crudeness, even the near-savagery, to which men were reduced in some portions of the frontier."[39] The nostalgic myth of the cowboy was suffering a collapse when Wallace Stegner wrote that the national icon of freedom, the cowboy, "was and is an overworked, underpaid hireling, almost as homeless and dispossessed as a modern crop worker, and his fabled independence was and is chiefly the privilege of quitting a job in order to go looking for another just as bad."[40] Stegner's voice gave this truth currency, yet he was not the first to notice. Edward Aveling, the son-in-law of Karl Marx, visited the United States in 1886: "Cowboys are a race exploited by ranch-owners as mercilessly as ever laborer was by capitalist. . . . The cowboy's life is not an ideal nor an idyllic one. . . . In a word, out in the fabled West the life of the 'free' cowboy is as much that of a slave as is the life of his Eastern brother, the Massachusetts mill-hand."[41]

Stegner and scholars such as environmental historians Donald Worster, Patricia Limerick, and Richard White initiated what came to be known as the school of New Western History and challenged Turner's idea that the West was a model for the American character. This group further argued that pushing back the frontier had not been as orderly as Turner claimed. It is now thought Turner's frontier thesis is more applicable to the Middle West and has little relevancy to either the Southwest or the American character. While the American character theory of Turner's "closing frontier" has now been largely discredited, the adventurous and land-hungry arrivals made an indelible mark, as much of the violence of the Old West lingered. Even though virtually all First Peoples' raids disappeared after the 1886 surrender of Geronimo, cattle rustling, false claims of theft, boundary disputes, and other conflicts kept violence close at hand. Writing in Silver City's *Desert Exposure*, Kara Naber quoted an early forest ranger, Henry Woodrow: "the Indians had quit killing people off, but these later settlers when they got tired of one another, the best man with a gun killed his neighbor and got him out of the way."[42] The Apache Wars were over, but this was still a threatening, rough country and the world in which Jeremiah McDonald and his sons endured. Pierce Rice's killing of Bart Irwin, the death of John Patterson and his assailant, and Bartley McDonald's killing of two young Hispanic men makes clear that this cruel ethos persisted well past the mid-twentieth century.

At this juncture the story of the ranch takes a sharp turn, from cattle ranch to restoration ranch. Although we still graze a few cattle, the focus is no longer beef and earning a living; rather our focus is habitat repair, a place for wildlife and at-risk species, science, and sequestration of the excess atmospheric carbon that is causing the climate and species extinction crisis. By the time we purchased the Pitchfork, the ranching and farming population nationwide had plummeted from 40 percent at the turn of the late 1800s to 2 percent, so low that the U.S. Census Bureau stopped counting the agriculture category, calling it "irrelevant." For the first time in the country's history, by 1996 more people lived in suburbs than in both rural and urban areas. This ever-increasing sprawl then moved beyond the suburbs, threatening not only ranching but all manner of wildlife that depend on uninhabited corridors and critical habitats suffering the fracturing onslaught of rangeland development. This change in land use has caused the loss of more than thirty million grassland acres—half to subdivisions and half to farm crops. Of the remaining grasslands, nearly half are depleted so badly that they are deemed beyond any potential restoration. In 2005, 16 percent of New Mexico ranchland sales were attributed to cattle and crop production and 84 percent to development, recreation, and other so-called amenity uses. Evaluating data from 1997 to 2017, a 2021 study for the New Mexico legislature by University of New Mexico's Bureau of Business and Economic Research found "the long-term trend of agricultural land, natural and working land loss has progressed unmitigated . . . [and] agricultural land conversion will accelerate over the next 20 years."[43] Fifty percent of ranches with federal permits are merely "hobby ranches," as working ranches have continued to disappear nationwide.

Before global overheating spiked, the Pitchfork Ranch typically received at least twelve inches of annual rainfall, but this has changed with periods of what some people carelessly describe as "extended drought." The ranch received less than ten inches of rain in each of the years from 2011 through 2021.

In 2012 and 2013, over three inches of rain fell within forty-eight hours, much of it draining off the ranch and leaving the habitat with as little as six inches of rain for the year. Recent years have been the driest in the last fifteen centuries on the North American continent. A 2022 *Nature Climate Change* study found the Southwest is in the worst drought in the past 1,200 years, estimated to be 42 percent attributable to the climate crisis. The global average temperature has been the warmest recorded since scientists began tracking these data in the late 1880s. The year 2021 was the hottest in human history. The current warming spike coincides with greenhouse gas buildup and dire predictions of global climate disruption from all reliable science organizations and credible climate scientists. A 2005 report painted a bleak picture for the American Southwest. When compared to the twentieth-century average, the study reported the American West had experienced an increase in average temperature during the previous five years that was 70 percent greater than the rest of the world.[44] Science of the climate crisis is clear: there is no credible support for hope of a return to historic rainfall or weather patterns. Permanently erratic, decreased rainfall here and around the planet has become carelessly considered the new normal of the climate crisis—the mass delusion of normalcy. There is nothing normal about this permanently increasing disastrous weather—better to refer to this worsening atmospheric process as "destructive weather," "deadly weather," or "lethal weather."

As we've seen with unprecedented heat waves, wildfires, floods, and superstorms, the intensity and frequency of extreme weather is continuing to worsen. The failure to recognize this near-unanimous scientific opinion mirrors the earlier failure to recognize John Wesley Powell's science-based truth about western settlement, although stakes now are much higher: a threat to organized global civilization.

Tipping points, with their unknowable and irreversible implications—such as the combined impact of widespread permafrost thaw, the end of crucial ocean currents, and the loss of Greenland's ice sheet—are decisive moments shrouded in the uncertainty of "when." The Pitchfork Ranch, although not exhausted beyond the ever-feared tipping point of degraded habitat, had long been on a downward trajectory toward a condition where restoration would have become economically futile. Deteriorating since European settlers' arrival, in 2003 the ranch was very different from the habitat Jeremiah McDonald began ranching over a century ago. Decades of carelessness left the ranch's land looking the same to the eye of a single lifetime, but from the perspective of several generations, the land has undergone dramatic change, all for the worse. The water table dropped, wells have gone dry, and the Burro Ciénaga was receding up-channel. The habitat was compromised for wildlife, as well as for humans.

The core goal of our ranch-wide restoration is to *slow the water and shallow the land so self-healing can begin*. We are arresting floodwaters in order to deposit eroded soil in the incised ciénaga and to slow water-sheet flow in the slopes and flats, thereby capturing sediment in low places and allowing water to seep and wick. We're working to return the ciénaga and surrounding land toward its condition before arrival of the Spanish. Installation of a variety of grade-control structures is helping the ciénaga and riparian watercourse reclaim themselves. The structures are also reconnecting surface and groundwater and will eventually allow the flood flows in the Burro Ciénaga to reach the terraces and their struggling vegetation. When the ciénaga's incision is filled with captured sediment, this historic arid-land wetland will be whole. A longtime rancher who has worked for more than fifty years restoring his ranch near Carrizozo, New Mexico, told us shortly after we bought the ranch, "You want to keep every drop of water that lands on your ranch to stay on the ranch."[45] We agree, and we've extended this idea to soil, with the objective of keeping every particle of sediment on the ranch and the soil that arrives with floods to remain here. Soil capture and retention are the goal.

As we began the restoration, we soon realized the work of restoring the ciénaga and surrounds not only serves this land and wildlife but also sequesters carbon dioxide, the primary greenhouse gas contributing to the climate crisis. This was a surprise discovery—an accidental treasure, the hidden gold—making the restoration of the Pitchfork Ranch wetlands a climate crisis sweet spot using nature to combat the climate, species extinction, and soil loss crises.

Although not readily apparent to the casual passers-by, the ranch is rich in natural, cultural, and historical bounty, layered with religious and cultural significance, environmental sensitivity, potential, and hope. It's clear from structure foundations and artifacts that Archaic people, Mimbreños, and finally the Apache people lived, farmed, and hunted here. This ranch rests in the heart of Apacheria, the Apache homeland and habitat that remains sacred to Apache people. One Chiricahua Apache origin story maintains they were birthed from these lands, meaning they were here first. A wildlife biologist who reintroduced the Gila topminnow here once told us, "The ranch is a microcosm. Its compact size allows one to get their hands and mind around the place with its amazing diversity, as well as a unique opportunity for science, research, and restoration."[46]

The habitat restoration and other work here is not about the right of proprietorship, title, or legacy; rather, it's about conservation, ethical caring, stewardship of place, and how each of us can play a meaningful role in the survival of life on the planet. After her second visit, a friend had this to say about the ranch:

> While the ranch doesn't attract world travelers and has somehow escaped the notice of writers who make a living describing unforgettable places, it riveted my attention the first time I saw it. It's a place I return to time and time again, in memory; a spot on this earth made beautiful by its testimony to a wild, harsh history and a tenacious refusal to give way to the elements—standing against decades of beatings delivered by

the poor stewardship of those who settled it and the varying climate changes of New Mexico's high desert.[47]

This ranch provides a platform from which to offer an account of the accelerating ecosystem collapse the world is facing. We live in different places with differing landscapes, creating a wide variety of ways to heal our wounded world. The ambition for survival requires us to do more to address these crises. Unless humankind takes serious measures to discontinue the causes of the planet's overheating, scientists warn that this emergency will end civilization as we know it. We have less than a decade—the Paris Climate Agreement gives us until 2030—to turn back these crises. If we don't get serious now, today's young and those yet unborn will inherit a world of unimaginable suffering.

CHAPTER TWO

Diverse Habitat, Wildlife, and Early Cultures

DIVERSITY MAKES THIS RANCH special: it holds one of the few remaining historic ciénagas, 8.3 riparian miles of the 48-mile-long Burro Ciénaga watercourse, history and myth, 34 Mimbres archaeological sites, rock art, mountains, slope land and grasslands, a section of the Butterfield Trail, all manner of plant and wildlife, natural beauty, and a small herd of Charolais cattle.

On the flip side, the ranch finds itself in a troubling condition, the desertified American Southwest linked to European arrival:

- The Southwest is the hottest region in the continental United States.
- The average New Mexico summer is 3.4 degrees warmer now than it was in 1984, and temperatures are projected to increase 5–7 degrees over the next fifty years.
- Up to 95 percent of ciénaga habitat has been lost since Spanish arrival.
- Riparian habitat makes up less than 2 percent of the American Southwest.
- Ninety-five percent of giant sacaton grassland has disappeared.
- Only around 30 percent of temperate grasslands remain healthy.
- Trees and shrubs now outcompete grass, flipping the historical botanical template from grassland to shrubland.
- Water wells throughout Grant County, where this ranch lies, are going dry, and all six of the Pitchfork Ranch wells have gone dry. New wells produce less than five gallons of water a minute.

The ranch lies at 5,100 feet elevation, just west of the Continental Divide in southwest New Mexico. Although mountainous, the land is primarily rolling Chihuahuan Desert grassland, one of the most biologically diverse arid regions in the world. The ranch is within the Priority Grassland Landscape evaluation by The Nature Conservancy (map 1). The conservancy's

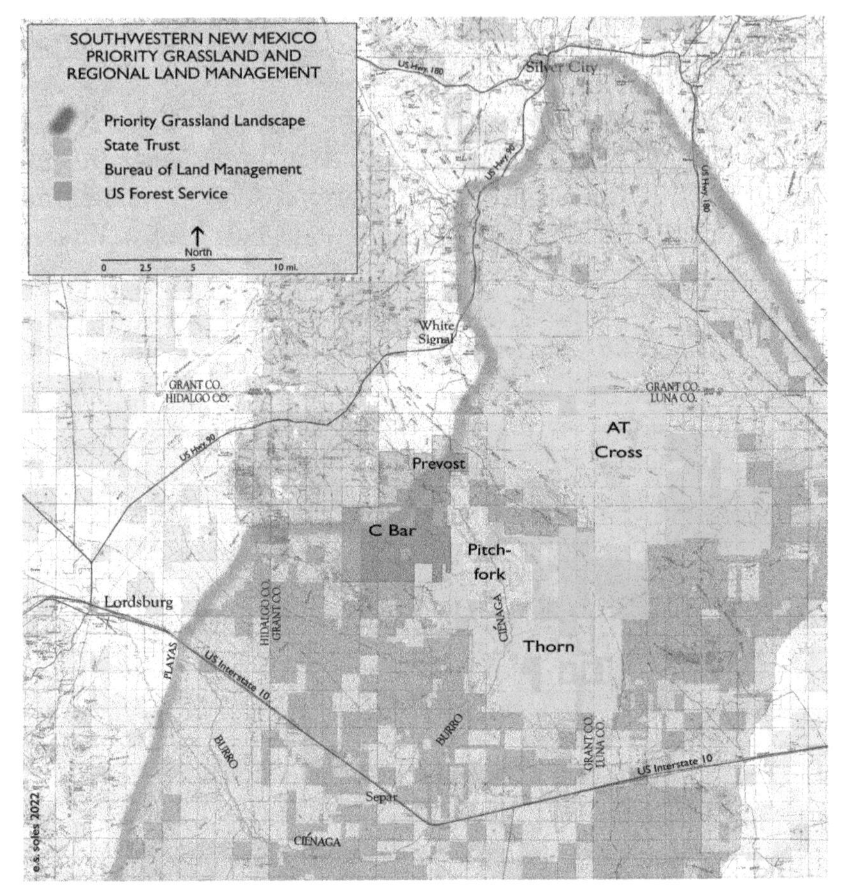

MAP 1 Three canyons converge to form the headwaters of the forty-eight-mile-long Burro Ciénaga at the northwest edge of the Otto Prevost Ranch. The alluvial fan or "cork" that creates the historic ciénaga—the key to its restoration—is one and a half miles from the Pitchfork Ranch's upper west boundary and speaks to the centrality of the Pitchfork for saving the ciénaga. Created by Ellen Soles after work of Steve Bassett of The Nature Conservancy, 2012.

assessment identifies a complex of priority grasslands where limited restoration investment will yield the greatest return in terms of restoring grassland health and recovering targeted wildlife species. Only about 19 percent of Chihuahuan grasslands remain dominated by native grasses and are relatively shrub free.[1] New Mexico has some of North America's best

FIGURE 5 The Chihuahuan Desert is home to a few riparian corridors, such as the one on the Pitchfork Ranch. Courtesy of A.T. Cole.

remaining grasslands, a portion of which are on this ranch. Grasslands are the most threatened habitat on Earth and the least protected, with less than 2 percent worldwide and just 4 percent in the United States formally safeguarded.

The restorable ciénaga is the ranch's most important feature. *Ciénaga* is a Spanish term used in the American Southwest for this rare desert wetland: marsh, a silty, spongy area; a bog; a shallow, slow-moving flow of water through dense surface vegetation; and permanently saturated soils in otherwise arid landscapes, historically occupying nearly the entire width of valleys (see appendix A). The 1958 *Cassell's Spanish Dictionary* defines *ciénaga* as a "marsh, bog, miry place."[2] We have a friend whose description of ciénagas is "wet meadows, occupying sites in the Southwest and perennially or near-perennially inundated by water with a substantial groundwater component."[3] They are found almost exclusively in the International Four Corners region of Arizona, New Mexico, Sonora, and Chihuahua. The phrase "arid-land ciénaga" distinguishes the spring-fed ciénagas in the Southwest from the saltwater ciénagas near oceans and a variety of other waters in South America. Commonly associated with springs, arid-land ciénagas usually occur where geomorphology forces water to the surface over large areas. Their shallow water is confined at the down-channel, lower end of the ciénaga by an alluvial fan of sediment that serves as a plug or "cork," holding back and pooling its water.

Ciénagas have their own unique characteristics, quite different from a river, creek, or stream. Historically, ciénagas were characterized by slow-moving, broad water flows through extensive, emergent vegetation. But today, the erosion accompanying post-settlement channelization has entrenched most ciénaga water between vertical walls. The drawdown of local water tables has dried up most marshland environments, leaving only remnant ciénagas, probably a mere 5 percent of what was here before the arrival of Europeans. Most remaining, damaged ciénagas look and function much like a creek: narrow, incised, continuing to deepen.

Before climate breakdown dominated other worries, the World Economic Forum's "Global Risks 2015" report warned that the global water crisis—lack of access to safe drinking water and extreme weather events—was the highest risk and the greatest impact on our lives and the planet. Research published in the journal *Science* in 2021 on "Cumulative Change in Biodiversity Facets" found rivers, lakes, and streams to be vital ecosystems covering less than 1 percent of the planet's surface. All but 14 percent of the world's river basins have suffered serious damage from human activity—the consequence of hundreds of years of so-called progress and carelessness. Climate and water scientist Brad Udall maintains "climate change is water change," meaning the most dire impacts of climate breakdown are felt in our rivers and clean water. These losses are critical because mangroves, seagrasses, marshes, and other wetlands like ciénagas absorb four to five times more carbon per hectare than forests. Wetlands conserve 20 percent of the world's organic carbon ecosystem while facing an average human-caused loss rate of 1 percent each year. Storing about five times more carbon dioxide than forests and as much as five hundred times more than oceans, wetlands are hotspots where conservation and restoration make an important contribution to the challenge of drawing down excess atmospheric carbon.[4] A local youngster got it right when she wrote, "Water has always been the difference between life and death, boom or bust, and it will be the difference once again between a sustainable future or no future at all."[5] Water sits at the foundation of everything.

In a global context, destruction of the few remaining ciénaga wetlands might seem trivial, but the loss of these sources of arid-land surface water has enormous importance to the endemic plants and animals that coevolved with and depend on these systems. Since the 1800s, natural wetlands and semiarid desert grasslands of the American Southwest have largely disappeared.[6] So too with soil. Before recognition of the escalating global climate crisis, there were those who maintained the loss and depletion of soil was the world's most threatening problem: "unless more immediate disasters do us in, how we address the twin problems of soil degradation and accelerated erosion will eventually determine the fate of modern civilization."[7] It takes so long for nature to create soil, particularly in the dry Southwest, that it's been suggested soil should be thought of as a nonrenewable resource.[8] It's priceless, an irreplaceable resource and essential to sustaining life on Earth.

Although we are working to restore the entire Pitchfork Ranch, there will always be an emphasis on returning the portion of the incised Burro Ciénaga to a fully functioning ciénaga, unique to the American Southwest and northern Mexico. There were many hundreds, if not thousands, of ciénagas in the Southwest before the arrival of Europeans, but upward of 95 percent are now estimated to be lost.[9] Until 2018, when 109 small, previously unknown and unstudied ciénagas were identified in New Mexico during the development of a state-sponsored Wetlands Action Plan for Arid-Land Spring Ciénegas, scientists had known of only 155 remaining ciénagas in the entire

International Four Corners region. Of those, 69 are dead; 18 are so severely damaged as to be beyond repair; 39 are fully functioning; and another 29, including the more than a mile-long portion of the Burro Ciénaga on the Pitchfork Ranch, have been deemed restorable. Because of the recognition and importance of ciénagas, the U.S. Geological Survey has begun studying them, and their numbers have spiked to as many as 475—threefold in a mere eight years since 2018, when 155 ciénagas were known—although they are so damaged that the "up to 95%" figure for loss of ciénaga habitat remains unchanged. In recognition of their importance, we are working to establish a Ciénaga National Monument with this 11,393-acre ranch and the 2,760-acre Otto Prevost Ranch north of us, with its 3.4 miles of authentic ciénaga and the headwaters of the Burro Ciénaga.

When we first visited the ranch in 2002, we didn't know what a ciénaga was; neither Lucinda nor I recall having ever seen

FIGURE 6 The headcut in the Burro Ciénaga, before a flood in 2005 carried away two football fields' worth of wildlife habitat. Photograph by Lucinda Cole, 2005.

or heard the term before. Although we had lived our entire lives in the Southwest's Sonoran Desert, we hadn't given the shortage of water much thought. We saw water in terms of a faucet or swimming pool. Shortly after moving to New Mexico, when we returned to the ranch after a short holiday in 2005, we were shocked to see that a major flood had carried away tons of the ranch's reach of the Burro Ciénaga watercourse soil and adjacent terraces. It started with a headcut (fig. 6). Headcuts are erosive features in a watercourse, slope wetlands, or drainages, where soil is eaten away and lost, leaving a vertical ledge in the bed of the waterway and creating a waterfall that undercuts, incises, and unzips the land up-channel. The flood removed wildlife habitat the length of two football fields, one-third of a field's width, and three feet deep. An entirely new channel—deeper and wider—was created in the area where the headcut had been several months earlier.

Cattle trails and old two-track wagon roads created shallow scars that led to this erosive phenomenon. Some restoration practitioners maintain that cattle trails and roads played a major role in creating most arroyos and the desertification of the Southwest. In arid lands, small nicks and divots fail to heal. They become incisions, incisions become trenches, trenches become gullies, and gullies became arroyos as the water leaves the range at an ever-accelerating speed. The 2005 erosive process that occurred during our vacation demonstrates the history of the Southwest, the ecology of which, as Aldo Leopold wrote, is "set on a hair-trigger" that will bust open and change in an instant. He understood that "early settlers did not expect this [erosive headcutting], on the ciénegas of central New Mexico, some even cut artificial gullies to hasten it. So subtle has been the progress that few people knew anything about it."[10] In the late 1880s, people in Tucson, Arizona, excavated a ditch to intercept the shallow groundwater and divert it for irrigation downstream onto a nearby floodplain. The diversion succeeded for several years until floods from heavy rains in the summer of 1890 cut into the ditch, creating a new arroyo and a fast-moving headcut that ripped out and carried away six miles and thousands of tons of soil in just three days. This process has been the norm for several centuries in the Southwest.

The need to accelerate the pace and extent of worldwide habitat restoration, particularly wetlands, which as peatlands, mangroves, salt marshes, and seagrass meadows cover only 1 percent of Earth's surface, yet store 20 percent of the global organic ecosystem carbon, is now undeniable. Although unknown in Leopold's time, recent research of wetland function has discovered a reciprocal organism-landform interaction called biogeomorphic feedbacks:

> Because carbon emissions from degraded wetlands are often sustained for centuries until all organic matter has been decomposed, conserving and restoring biogeomorphic wetlands must be part of global climate solutions. . . . [This] highlight[s] the urgency to stop through conservation ongoing losses and to reestablish landscape-forming feedbacks through restoration innovations that recover the role of biogeomorphic wetlands as the world's biotic carbon hotspots.[11]

The natural functioning of wetlands is a process with far greater carbon capturing capacity than previously known, increasing the importance of wetland preservation—we're destroying 1 percent of Earth's wetlands each year. Restoration is the only way to maintain what we have and reverse losses. Leopold's advocacy for habitat restoration focused on habitat and wildlife: "A rare bird or flower need remain no rarer than the people willing to venture their skill in *building it a habitat.*"[12] Although right for its time, increasing global temperatures render this view too narrow. Restoring ciénagas and other wetland habitats is critical, not only because of their carbon capturing capacity, but because they are the kidneys of river systems, the most vulnerable of Earth's habitats.

Rivers and other freshwater systems have the highest extinction rate per unit area in what is already a water-scarce world.

The 2019 Intergovernmental Science-Policy Platform on Biodiversity and Ecosystem Services' global assessment of ecosystems and wild populations found that 85 percent of the world's wetlands have been lost. With the threat of climate overheating, wet spots have become critical ecosystems in arid environments. They have taken on increased importance, acting as carbon sinks, drawing carbon dioxide from the planet's overheated atmosphere back into Earth. Carbon sinks are one area of the climate emergency question about which there is little dispute yet almost no discussion. Environmental literature is replete with warnings that Earth's capacity to absorb the filthy byproducts of global capitalism is exhausted. This warning was made almost fifty years ago by Donella Meadows, Jørgen Randers, Dennis Meadows, and William W. Behrens III in *The Limits to Growth*, at one point the world's most widely read environmental book. Although photosynthesis is infrequently mentioned and not thought of as a leading candidate as a climate crisis solution, it clearly should be, and soon will be, once the importance of natural climate solutions is recognized and incorporated into the suite of fixes to the climate, species extinction, and soil loss crises.

The United States has lost over 50 percent of its wetlands, while in seven states over 80 percent of original wetlands have disappeared. The situation is worse in the Southwest, with only a thin 5 percent of ciénaga habitat remaining. Our restoration initiated the repair of the Burro Ciénaga's damage with the installation of grade-control structures that capture suspended sediment in flood flows, thereby raising the channel bed, increasing vegetation and groundwater, and in due course allowing floodwater to course outside of what has become a deeply incised channel. This restoration will eventually recreate the ciénaga wetlands and giant sacaton flats that once covered the valley. While that goal remains central to our work, we have come to realize the crucial corollary benefit of the potential to sequester atmospheric carbon as a natural climate solution.

A number of how-to publications detail the variety of grade-control structures—often referred to as beaver dam analogs, rock detention structures, erosion-control structures, grade-stabilization structures, natural infrastructure in dryland streams, or, as in a recent *Science News* article, Zeedyk structures—and techniques used for sediment retention and restoring damaged habitat in the Southwest. *Let the Water Do the Work* by Bill Zeedyk and Van Clothier is the book and restoration protocol Lucinda and I rely on and see as best for addressing damage to arid Southwest terrain. There is no need to survey the techniques or construction methods for restoration structures; Zeedyk has done that. Although I'll leave the how-to to him, in chapter 6, photographs show the changes on the ranch that have resulted from the installation of erosion-control structures.

The ranch has benefited from twenty government grants and the labor of volunteer groups that subsidize the ongoing landscape-wide restoration. The work has resulted in the installation of more than 200 grade-control structures in the main channel and more than 800 structures into 32 of the 33 side drainages that empty into the reach of the Burro Ciénaga on the ranch. These erosion-control structures have captured thousands of tons of sediment, raised the groundwater table—eleven inches at one point, although the shortage of rain may have cut back these gains—and aggraded the channel nearly two feet throughout and five feet in some areas. We've also installed a water catchment with two 1,800-gallon holding tanks, with seven more scheduled, that capture rain, providing wildlife with water throughout the year. In 2013 the State of New Mexico funded what had been our most substantial grant: a side-channel project requiring the purchase of 683 tons of two-foot boulders for machine-built structures in the larger drainages and an untold number of on-site rock structures for smaller drainages. A similar 2023 state grant more than doubled the size of the 2013 grant, subsidizing installation of grade-control structures in the lower three miles of the

8.3 miles of the Burro Ciénaga in the Pitchfork, which backs up or slows water up-channel.

Restoring the staggering amount of damaged habitat in the Southwest and elsewhere on the planet is imperative for the future of civilization; restoring habitat that is at risk of outright elimination, like ciénagas, is an emergency. Although the power of the Endangered Species Act (ESA) rests on its mandate to protect habitat critical for the recovery and maintenance of a species, the act does not yet protect habitats independent of individual species. If it did, with up to 95 percent of ciénaga habitat lost, they would surely qualify for protectorship. Ciénaga restoration is essential for endemic species preservation. Scientists are unanimous that we are in the midst of the planet's sixth major extinction event. Ciénaga restoration is part of the solution to this massive loss of species, not only because of the animals and plants unique to them, but also because ciénagas "have the potential to represent a great success story in conservation given that the degradation of these systems is relatively recent and we observe a great resilience in ciénega vegetation released from disturbance pressure."[13]

Restoration potential is also significant in view of the ecosystem services (nature's benefits that serve humans, heretofore little recognized) that ciénagas provide when functioning properly. Yet in the face of the climate and companion crises and despite the potential of natural climate solutions to reduce atmospheric carbon, land-based carbon-sequestration efforts receive only about 2.5 percent of climate mitigation dollars.[14]

Thousands of tons of topsoil and other valuable habitat components have eroded from the Burro Ciénaga watershed over the last two hundred years. Before we began restoration, the vertical walls of the watercourse were becoming larger and the incision deepening, gullies were eroding into ever larger arroyos, and the overall landscape was being stripped of soil so severely that areas were simply barren rock surfaces, devoid of soil, grass, and other life. There was a time when this didn't seem to matter. A sense of how poorly ciénagas were viewed historically can be gleaned from a remark made during the naming of Silver City, New Mexico. When city fathers met in 1870 to choose a name for the community occupying the once unmolested ciénaga, a lengthy discussion finally reached consensus to discard La Cienega San Vicente and call their new town Silver City. After reaching a consensus, one of the men remarked, "It was one hell of a name to call a town on a mud flat."[15]

Another example of the historical indifference to ciénagas is seen in this dialogue from the beloved New Mexico novel *Red Sky at Morning*. In a 1940s conversation between the novel's narrator, Joshua Arnold, and his classmate, this exchange occurred:

> "I didn't know there was this much water around Sagrado," I said. "The Sagrado River's been dry since I got here."
>
> "This is a *cienega*," Parker said. "It's some kind of underground spring, but it's not good for anything but making the ground wet. Costs a fortune to drain it or pump it off, and Cloyd isn't about to spend money for things like that."
>
> "Does the whole family have to walk through this stuff to get to the house? You'd think they'd build a duckwalk or something." Parker said Cloyd would rather swim than build something useful.[16]

In 1985 two ichthyologists, professor W. L. Minckley and his student Dean Hendrickson, published a paper that alerted the academic community to the importance of the Southwest's overlooked ciénagas. They suggested these dwindling, invaluable, yet little-understood ecosystems were a resource meriting further study.[17] The paper is credited for creating long-overdue interest in this unique desert wetland and may also have led to inclusion of *ciénaga* in the climate crisis lexicon. In most regions in America, "the accumulation of soil and water created an estuarine paradise swarming with waterfowl, crustaceans, fish, and a whole assortment of swamp creatures that scared the hell out of early settlers and convinced them that the only good wetland was a drained wetland."[18] Yet the importance of wetlands and ciénagas in the Southwest is staggering: wetlands are the keystone ecosystems in arid environments, historically comprising only a tiny percent of the surface area of the arid Southwest—now only a sliver. Simply by increasing the wetness in otherwise arid regions, desert ciénagas and riparian corridors like the waterway on the Pitchfork Ranch can increase regional diversity by as much as 50 percent.[19]

Scientists studying soils—in the form of cores drawn from ciénagas—have begun the task of teasing out the natural processes that established ciénagas and other landscapes in the Southwest over the last eight thousand years. In the process, they have discovered the details of what caused the loss of ciénagas. In less than two hundred years, a disconnected series of mostly human-caused activities transformed ciénagas and the landscape of the Southwest from a *depositional* habitat to an *erosional* environment, from braided wetlands to single-channel dispersal, thereby severely lowering groundwater tables, causing the desertification of the landscape, and resulting in thousands of acres of ciénaga habitat losses.

Arguably the most significant cause of ciénaga losses was the dewatering of the Southwest by the over-trapping of beaver that began in the 1820s. *The Personal Narrative of James O. Pattie* is among the earliest accounts of this assault on the wildlife of the Southwest and provides a detailed—though much disputed—tale of Pattie's exploits among First Peoples and Spanish settlers between 1825 and 1830. The party killed every beaver they could find, moved on, and did the same wherever they went. In his history of New Mexico's Gila River, Gregory McNamee notes that "with the demise of the beaver came a major alteration of the Southwest landscape . . . and for the first time, erosion became a major problem as [waters] flooded unchecked. Animal populations fluctuated wildly as their habitats began to disappear."[20]

Discussing the assault on the West by the cattle corporations, fur companies, and mining companies and their get-rich, absentee-owners, some from England and Scotland, Stegner asks: "Who among the mountain men would have paused to consider, or would have cared, that beaver were a water resource, and that beaver engineering was of great importance in the maintenance of stream flow and the prevention of floods?"[21] These giant rodents can build up to twenty dams per mile of stream, smearing water flows across the landscape.[22] Beaver dam building was a region-wide process of arresting water flows that transformed torrents into a series of wetlands and pools connected to a myriad of shallow, multiply branched channels.[23] It's been suggested that beavers shaped the North American continent.

The loss of these quintessential "ecosystem engineers" hastened a cascading process, compounded by later human misuse, resulting in ever-worsening, accelerated erosion and near elimination of ciénagas. These losses in turn accelerated the conversion of dynamic and complex stream and river ecosystems into today's simplified water delivery systems, pervasive throughout arid lands. The Southwest was a far different place from the one we know now. Although it's unimaginable in the present climate, "people could walk in the shade from the headwaters of the Gila River in New Mexico all the way to Yuma," Arizona, where the Gila empties into the Colorado River.[24] Restoration practitioners, trying to return habitat to its "pre-settlement condition," are often chided with the claim that no one knows what the pre-settlement condition looked like. Yet Kent Woodruff, Methow Beaver Project coordinator, believes beavers are "landscape miracle drugs" and adds, "We're not smart enough to know what a fully functional ecosystem looks like. But [beaver] are."[25] The grade-control structures that we install on the ranch and others that have been placed throughout the Southwest mimic the work of these original ecosystem engineers.

The damage caused by the virtual elimination of beavers may have been the worst, though not the first, major step in the dewatering of the Southwest. Ciénaga sites in an otherwise arid landscape provided the earliest water used by the first Spanish for their livestock. They brought increasing numbers of sheep into Mexico beginning in 1521; then Coronado brought even more into what would become the American Southwest in 1540. By 1586 there were thousands of sheep in the Southwest, increasing in number through the 1700s. Horses, goats, mules, pigs, burros, and sheep accompanying Juan de Oñate's caravan laid the foundation for the livestock industry that would dominate the region until it was converted and narrowed with the arrival of cattle.[26] As early as the 1680s, Pima First Peoples in Sonora complained that Spanish livestock were so thick that watering places were drying. By the late 1700s, sheep were a major regional industry, outnumbering cattle thirty-seven to one. But with the arrival of the Southern Pacific Railroad, the spread of the windmill, and the elimination of the Apaches, cattle ranching exploded between 1881 and 1884. By 1884 overstocking mushroomed: "every running stream and permanent spring had been claimed and adjacent ranges stocked with cattle . . . and by 1890 the entire region must have looked like one big cattle ranch."[27] The "1880s through early 1900s represented a period of grazing at extreme intensities greatly over carrying capacity of the plant communities."[28] Cattle numbers peaked in 1890, and the 1891 to 1893 drought resulted in a huge decline, yet the pattern of overstocking persisted.

Throughout the Southwest, the effects of long-term, sustained grazing of livestock—particularly during drought episodes—has reduced plant cover and plant vigor, resulting in the loss of biomass, and has increased runoff and sediment transport, thereby worsening the desertification process. Overstocking led to the formation of deeply incised channels, water table declines, and further loss of wetland vegetation, making the Southwest's ciénagas among the most abused sites on Earth. In his comprehensive study on the impact of humans and their cattle corporations on the Arizona borderlands in *A Legacy of Change*, Conrad Joseph Bahre concludes,

> Probably no single land use has had a greater effect on the vegetation of southwestern Arizona or has led to more changes in the landscape than livestock grazing and range

> management programs. [Cattle] grazing since the 1870s has led to soil erosion, destruction of those plants most palatable to livestock, changes in regional fire ecology, the spread of both native and alien plants, and changes in the age structure of evergreen woodlands and riparian forests.[29]

Donald Worster writes that the collapse of the grasslands of the West between 1880 and the end of the decade due to overstocking "was one of the greatest, as measured in the loss of animal life, in the entire history of pastoralism."[30] The first comprehensive effort to measure the impact of cattle and sheep in the West came in 1936 in the form of a 620-page government report titled *The Western Range*, which covers the period from the first days of cattle ranching to the Dust Bowl and concludes, "There is perhaps no darker chapter nor greater tragedy in the history of land occupancy and use in the United States than the story of the western range."[31]

Because grasslands represent the most threatened land on Earth and the least protected, there is increasing recognition of these losses and efforts to implement protection for this important part of our world. The Nature Conservancy has identified the Pitchfork Ranch as among some of North America's largest and best remaining desert grasslands.

Though its normalcy went unrecognized by pioneering newcomers, the arid Southwest's uneven, severe weather and the drought of the late 1880s and early 1890s exacerbated the absence of fire, loss of the beaver, and the overstocked landscape that had already severely degraded grass and wetlands. After an unusually dry summer in 1886, the consistently below-zero temperatures in the following winter were so bitter that cattlemen could not fathom the late spring that followed. The weather worsened, and livestock mortality rates reached 75 percent.

In 1901, with cattle numbers greatly reduced as a result of the drought, D. A. Griffiths, chief botanist in charge of grass and forage plant investigations for the Arizona Experiment Station in Tucson, Arizona, set out to determine what the range conditions were before the livestock boom of the 1880s. Apparently unaware of the role of beaver, he asked two men who had lived through that period to attribute the present unproductive condition of the range to cattle overstocking, drought, or both. Their answers: cattle overstocking and beaver. Griffiths learned from Oracle, Arizona, rancher C. H. Bayless: "About twelve years ago the San Pedro Valley consisted of a narrow strip of sub irrigated and very fertile lands. Beaver dams checked the flow of water and prevented the cutting of a channel. Trappers exterminated the beavers, and less grass on the hillsides permitted greater erosion, so that within four or five years channel varying in depth from 3 to 20 feet was cut almost the whole length of the river."[32] Springs that were long thought to be permanently wetted went entirely dry. The 1878 United States Geological Survey report by John Wesley Powell warned the shortgrass plains were ill suited for intensive grazing because of recurrent droughts and so little rain west of the hundredth meridian. But nobody listened: house-high piles of cattle bones followed, and a severely damaged Southwest remains today.

Ciénagas also suffered damage when pioneers recontoured the broad ciénaga canyon flats in a misguided effort to prevent flooding of their recently established agricultural fields. Remnant ditches, dikes, and dams remain around abandoned agricultural fields near many of today's few remaining and poorly functioning ciénagas. While most farming has long since ceased, the resulting channelization and concentrated flows have reduced these historic wetlands to a fraction of their original size and created deep, vertically walled incisions that have progressively worsened and lowered the groundwater table, further dewatering former ciénaga wetlands. While today the ciénaga is a Goodding's willow–lined, incised channel 10 to 25 feet wide and 5 to 15 feet deep, before the arrival of Europeans, the ciénaga was a shallow marshland several football fields wide reaching the toes of the mountains on both the left and right in figure 7. We encountered this narrowed and incised creek-like ciénaga when the restoration began. For

FIGURE 7 The current ciénaga along the tree line (*center*), with blue lines denoting the former ciénaga area that will be revived when restoration is complete. Down-channel left of the tree line, on the right in the photograph, are two abandoned ten-acre agricultural fields. Courtesy of A. T. Cole.

eons, ciénaga waters flowed across this land before being channelized by a two-foot dike that ran from the toe of the mountains on both sides of the valley. Cutting the ciénaga in half and diverting water at the fifty-yard line or midfield created an incision that continually deepened until we began restoration. The berm has been removed.

Fire suppression also contributed to the dewatering of ciénagas and the desertification of the Southwest. The consequences of eliminating fire in the forests, the dangerous buildup of plant fuel, and the overheated climate that has led to the current spate of catastrophic forest and brush fires is fully understood, although the implications of fire suppression for the health of the open range and ciénagas are less recognized. Fire was a significant factor in the evolution of ecosystems in the Southwest, and many plant species adapted to fire as a result of lightning strikes and deliberate ignition by First Peoples. A number of studies in multiple disciplines makes clear that southern Arizona vegetation was frequently burned before European arrival.[33]

Although historic records mention deliberate and accidental burning of valley vegetation by the new residents, naturally occurring fires have been suppressed to such an extent since Spanish arrival that woody plants have outcompeted grasses and transformed historic grasslands into a landscape of trees and shrubs. The majority of grass biomass is found in their root systems that lie beneath the surface, whereas most of the biomass of trees and other woody plants is just the opposite—above ground. Fire helps grass and harms trees. Most tree-ring studies of fire in the region have also concluded that fires have reduced in frequency.[34] Studies of microscopic charcoal, pollen, stable isotopes, and elemental fractions of organic material drawn from ciénaga cores correspond with tree-ring evidence that the incidence of fires decreased in the Sonoran

FIGURE 8 The site of a fire started by lightning on a hill near the ranch, killing the trees on the right before it was brought under control by the local fire department. Grass persists. Photograph by the author.

Desert two hundred years ago with the displacement of First Peoples' agriculture by European settlement.

Before the arrival of Europeans, Southwest grassland fires occurred on average every eight to ten years and killed many woody plant species, but merely topped off and actually strengthened the health of grasses. Before woody plants encroached so severely onto these grasslands, abundant grasses regularly arrested large quantities of sediment. Expansive ciénaga surfaces dispersed floods into sheet flows, helping prevent the channelization so ubiquitous today. Floods erode barren soils and turn what had been shallow runoff into gully gushers, worsening the already massive incisions and arroyos throughout the Southwest. The result is heightened flash flooding and channel scouring, lowering water tables and contributing to the dewatering of ciénagas.[35] The dominant land surface process in the Southwest now is scouring, the opposite of slow-moving water being backed up behind beaver dams, which resulted in abundant grass and other components of the arid-land habitat that existed before the arrival of Europeans.[36]

Springs and ciénaga ecosystems are among the most structurally complicated, ecologically and biologically diverse, productive, evolutionarily provocative, and threatened ecosystems on the planet.[37] As virtual oases, ciénagas exert vastly disproportionate positive impacts on regional ecology, evolutionary processes, and sociocultural economics in relation to their size, along with providing critical habitat for numerous endangered plants and animals. Functioning as aquatic archipelagos, ciénagas not only harbor unique and endangered aquatic organisms but often contain a variety of important fossil remains of prehistoric and long-lost animals—mammoths and mastodons among them. Given the small land areas ciénagas occupy, their role as fertile habitat, migratory bird rest stops, and biodiversity centers in arid regions is second to none.[38] Scientists think of ciénagas as Noah boats, ark-like habitats carrying flora, fauna, and important artifacts from the deep past: ancient pollen, stable isotopes, archaeological leavings—nature's archive—and a unique climax ecological community of significant biotic value. The 95 percent loss of ciénaga habitats, their importance as keystone ecosystems, and their corollary benefits render arid-land ciénagas shoo-in candidates for top-priority restoration efforts such as the work being pursued on the Pitchfork Ranch.

The abundance of plant and animal life on the Pitchfork Ranch has been thoroughly documented by a number of scholars and volunteers. The animal inventory consists of 31 mammals: mule deer, elk, pronghorn (antelope), coyote, 8 species of mice, black bear, badger, cougar, 4 species of rat, chipmunks, rock squirrel, northern raccoon, collared peccary (javelina), 3 species of skunk, white-nosed coati, bobcat, ringtail cat, several species of bat, black-tailed jackrabbit, and on one occasion, a wolf. It also includes 33 amphibians, reptiles (3 varies of rattlesnake: coontail, blacktail, and rock rattler), a fish, 153 birds (21 of which are considered by Partners in Flight to be species of continental importance in the North American Landbird Conservation Plan of 2004), 59 species of butterfly, 563 species of moth, and 6 mosses.[39]

Before we moved to the ranch, when Professor Emeritus Dale A. Zimmerman, co-author of *Birds of Kenya and Northern Tanzania*, was teaching at Western New Mexico University, he would bring his students to the ranch for study and knew the birds here. We commissioned a painting from him of birds we have seen from our front porch (fig. 9). Dale and his wife, Marian, visited the ranch, and our conversations informed us of the reordering in bird populations as a result of the changes in climate, a phenomenon that perplexed him until the implications of the overheated climate became ubiquitous.

The ranch hosts 342 species of plants, 64 of which are grasses. There are 5 expanding giant sacaton fields of 10 or more acres as well as sacaton terraces throughout the ranch's reach of the Burro Ciénaga. The Grant County Native Plant Society and representatives of the Jornada Experimental Range have collaborated on a plant inventory that continues to add species,

FIGURE 9 Dale Zimmerman's watercolor painting of the birds of the Pitchfork Ranch.

not simply due to a thoroughness of ongoing surveys, but also because new varieties show up as a result of restoration-improved habitat. Just one example, the cardinal flower, arrived on the ranch in 2013.

The warming climate likely plays a part in fifty-five species of terrestrial mammals and birds having increased their breeding distributions northward into the American Southwest since 1890. Several of these animals are common on the Pitchfork Ranch: Lucy's warbler, cardinal, hooded oriole, and white-nosed coati. The northernmost New Mexico sightings for the northern pygmy mouse (captured by a student from Western New Mexico University in Silver City), *Horama zapata* moth, and Botteri's sparrow were identified on the Pitchfork Ranch. These sightings reflect the notion that not only are species moving to cooler habitat—either northward, to higher ground, or to deeper waters—but habitat is changing too. Changes in animal distribution suggest recent modification in landscape character, that is, the transition in vegetation from open grassland and woodlands to more closed communities of forest and scrubland.[40]

In addition to its complex ecological composition, the ranch's cultural history and associated cultural material are extensive. A storage cave from the Archaic period with a small wooden overhead beam sealed in mud is well preserved today. In April 2007 a dozen arrowheads were found on a mound adjacent to the Burro Ciénaga; one was a Clovis point, as much as eleven thousand years old. Because of safety, access to water, and perhaps beauty, successive peoples inhabited these elevated areas near water, leaving mixed materials covering many thousands of years.

Between 1000 and 1130 CE, the Classic Mimbres period, as many as 6,000 farm-based Mimbreños people lived along the 91-mile-long Mimbres River, 16 miles northeast of the Pitchfork Ranch. Most of the cultural material found here belonged to those First Peoples. Archaeologists from Archaeology Southwest in Tucson have identified 34 Mimbres sites along the ranch's reach of the Burro Ciénaga, including one 5.6-acre site surveyed by archaeologist Patricia Gilman of the University of Oklahoma and found to have been occupied continuously from 750 CE to 1130 CE.

The Puebloans are the often-overlooked link between the First Peoples who dispersed—the Mimbres, Hohokam, and the Ancestral Pueblo (Anasazi) people—due to adverse environmental conditions and the arrival of the first Europeans from

FIGURE 10 Pitchfork Ranch artifacts. The large, smooth stone is a work bench or lapstone. *Left to right*: black Clovis point, possibly as old as eleven thousand years (possibly traded and brought here or evidence of habitation during that time); a white stone drill (possibly for bead-making or similar activities); a textured pottery sherd with indentations made by a fingertip; and two Classic interior pieces of a ceramic pottery bowl. Two additional pottery sherds and a Mimbres projectile point; a stone weight for an atlatl (a device for throwing a spear, precursor to the bow and arrow) with a carved groove for twine attachment; two white stone bird points; and two turquoise beads. *Below lapstone, left*: a mano, used to grind grains on a metate or milling stone. (Note the markings. Our thinking is that it is a signature indicating ownership. I can almost hear the woman pointing to the marks, saying, "This one is mine.") *Below lapstone, center*: a stone carving we believe to be an effigy of either a bear or a turtle or a combination of both that we refer to as a "beartle." *Below lapstone, right*: a handheld scraping tool used to skin animals, carve, or cut meat.Photograph by Lucinda Cole, 2016.

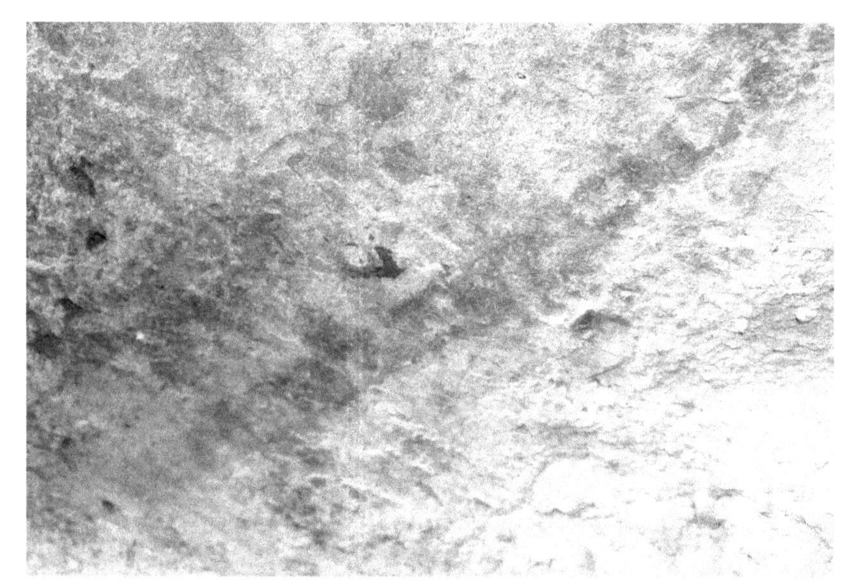

FIGURE 11 A mountain spirit painted by the Red Paint People. Photograph by Lucinda Cole, 2010.

Spain. After the dramatic changes brought on by the warming of the 1100s, people dispersed, adapted, pursued a new way of living, and made their home in the Southwest for four centuries before the Spanish began exploring these long-inhabited lands in the 1500s and settling here in 1598. Before the Spanish entrada, the American Southwest witnessed an industrious and occasionally turbulent period of significant reorganization. Not to be confused with the current human-caused overheating of the planet, the period between 1130 and 1540 CE saw major climate shifts, extensive drought, and the Little Ice Age that triggered movement and instability in search of suddenly scarcer resources. The adjustments these First Peoples made suggests the adjustments we must make to the crises we are experiencing.

In the heart of Apacheria, this land was dominated by the Apaches for several centuries before the surrender of Geronimo in 1886. Paintings by the Red Paint People remain under mountain overhangs on the ranch. There are a dozen pictographs at an isolated location on the north portion of the ranch. We understood one to be a painting of a hellgrammite, the aquatic larva for the dobsonfly, commonly used as fish bait. However, several years ago I showed the photograph (fig. 11) to a Chiricahua Apache elder in Silver City, and he expressed surprise when I referred to it as a hellgrammite and corrected me: "Oh no, that's a mountain spirit."

A U.S. Army heliograph station was located on the south portion of the ranch, where the U.S. military used "talking mirrors" and Morse code to counter Apache smoke signals, and the U.S. Cavalry's Camp Henley was on the north portion in 1886. There is also an Apache safe-haven hideout in a high, isolated mountain area just off the ranch. Born in the Middle Fork of the nearby Gila River, Geronimo likely passed this way when walking back to his homeland from the 1851 Janos Massacre, where his mother, wife, two or three children, and more than a dozen other women and children were massacred by angry Whites from Tucson.

FIGURE 12 Cooper's Doubt in the perennial reach of the Burro Ciénaga in the fall after initial phase of restoration. The grade-control structure is visible at right. Photograph by Dennis O'Keefe, November 2008.

Far more important than often credited, the ranch is teeming with beauty, most apparent where the permanent water courses its way southward. Beauty was one of Aldo Leopold's principal focuses and woven throughout his writing: "Bread and beauty grow best together. There harmonious integration can make farming not only a business, but an art; the land not only a food-factory but an instrument of self-expression, on which each can play music of his own choosing."[41] If we are to embrace Leopold's land values, beauty is essential. Despite the ciénaga persisting as a creek-like shadow of its historic self, it remains beautiful.

The beauty of the Burro Ciénaga (fig. 12) was once far different from its condition today due to new sediment aggregation and abundant vegetation. A post-baffle grade-control structure—a row of hand-driven posts creating a meander that slows the water, pushes it leftward, increases wicking, and captures suspended sediment—is now a foot below the water surface. The watercourse looks and functions less like a creek, still many restoration years away from a fully functioning ciénaga.

FIGURE 13 Horizontal yellow, Soldier's Farewell Hill, from the north portion of the ranch. Photograph by Alicia Arcidiacono, November 2017.

Soldier's Farewell Hill is the ranch's most dominant, historically well-known feature and is a region-wide landmark. The Navajos referred to this region with a term that translates to "horizontal yellow" (fig. 13).

Both scholars and the public have given short shrift to the important ideal of beauty. Beyond painting, music, and the arts, the idea of beauty has been relegated to a matter of artistic taste, not science. Yet its recognition is important to the task at hand. Charles Darwin gave beauty high billing—on par with and equal to natural selection—in his "aesthetic theory . . . of mate choice." Richard O. Prum's *The Evolution of Beauty*, maintains Darwin's writing is clear in its recognition that "natural selection alone cannot possibly explain the diversity, complexity and extremity of the sexual ornaments we see in nature."[42] Before unearthing the importance of beauty, Darwin famously wrote to his friend and botanist Asa Gray in Concord, Massachusetts, "The sight of a feather in the peacock's tail, whenever I gaze it, makes me sick."[43] Having no apparent survival value, the extravagant feathers conflicted with what he had written in his *On the Origin of Species by Means of Natural Selection*. Sexual ornaments can "not only fail to signal anything about objective mate quality, but can actually lower survival and fecundity of the signaler and the chooser." Once Darwin understood the role of beauty in evolution, in his next book, *The Descent of Man, and Selection in Relation to Sex*, he hypothesized how female preferences were a "powerful and independent force in the evolution of biological diversity."[44] Prum explains how Victorian values steeped in patriarchy account for beauty being given meager merit in evolutionary biology. What's important for our purposes is to recognize, respect, and prioritize beauty, not as a mere aesthetic of art, design, or femininity, but rather as an essential element of our being, science, and land ethic. After all, it's increasingly apparent we are hardwired to require beauty as much as food and shelter.

J. Baird Callicott's essay "The Land Aesthetic," in his *Companion to "A Sand County Almanac,"* chronicles the shortcomings in the writings of many who followed Leopold's lead in pursuit of a land ethic. He maintains we've ignored Leopold's reliance on beauty in the original and revolutionary "land aesthetic" that persists throughout *A Sand County Almanac*. Leopold's interest in beauty was as important to him as a proper ethical attitude.[45] In Western thought prior to the seventeenth century, beauty in nature was simply a source of aesthetic experience, and, Callicott suggests, it enjoys little currency in environmental and ethical considerations today: "The land aesthetic, desultorily and intermittently developed in *A Sand County Almanac* . . . is a new natural aesthetic, the first, to my knowledge, to be informed by ecological and evolutionary history and thus, perhaps, the only genuinely autonomous natural aesthetic in Western philosophical literature. . . . The land aesthetic enables us to mine the hidden reaches of the ordinary; it ennobles the commonplace; it brings natural beauty literally home from the hills."[46]

Rapid degradation of watersheds across the nation was Aldo Leopold's abiding concern and brought him to confront the universality of the challenges facing the protection of important habitat. In an unpublished, untitled manuscript, he wrote: "The government cannot buy 'everywhere.' The private landowner *must* enter the picture. . . . The basic problem is *to induce the private landowner to conserve his own land*, and no conceivable millions or billions for public land purchase can alter that fact, nor the fact that so far, he hasn't done it."[47]

Another important aspect of the ranch story is rain. The Pitchfork Ranch's rain history corresponds with what climate scientists have found throughout the world: overheated climate equals dramatic changes in weather. Our rain days measuring .06 inches or less averaged 10 in number for our earlier years here but doubled over the 10 years before this writing. Although we are now experiencing more rain days per year—52 versus 35, on average—they are often light rains or faint sprinkles with less water soaking in. More light rain days mean less absorption and more evaporation. At the same time, heavier rains have also increased; these rains result in more runoff, meaning we can't, functionally speaking, include all of this excess rain in our annual rainfall estimates because the runoff doesn't benefit vegetation. Several years ago, we started noticing one-seed junipers dying in the upland. The numbers of dead trees have increased, year after year, currently at 10 percent of total juniper trees, a function of warmer weather and the shortage of rainfall. Due primarily to the absence of fire, like much of the Southwest, the last several centuries have seen the northern portion of the ranch suffer increasing numbers of one-seed juniper trees encroaching on what was the horizontal yellow of grasslands. The long-settled infestation of this woody plant appears to be at an end, as they cannot persist in this land of lesser rain.

Speaking of trees—typical of how the ranch keeps on giving—in early 2023, Dr. Laura M. Norman and her crew from the U.S. Geological Survey office in Tucson visited to inspect our restoration as part of their studies on grade-control structures and their impact on erosion, water flow and retention and carbon sequestration. In a follow-up email she reported:

> The abundance of pristine Emory oak at your place is AMAZING.
>
> Acorns from the Emory oak tree are a critically important resource for the Western Apache tribal nations—including the Yavapai-Apache, Tonto Apache, San Carlos Apache and White Mountain Apache in east and central Arizona—who use it both for food and cultural and ceremonial purposes. Groves of Emory oak have been declining in health and yielding fewer acorns with each harvest for several decades due to loss of habitat, fire suppression, livestock grazing, groundwater reductions, species competition and climate change.[48]

CHAPTER THREE

Cattle

RESTORING LAND AND OTHER natural climate solutions are among the most important elements of the climate-survival agenda. Support of the western livestock industry's continued viability, as part of that program, is one of the main ideas of this book. Yet, if the world's cattle were a country, that country would be the third-largest greenhouse gas emitter behind China and the United States. The cattle question is central to ranching and has serious implications for the planetary climate. Animal agriculture and meat-eating accounts for about 15 percent of greenhouse gas emissions, falling fourth behind burning fossil fuels, transportation, and manufacturing.

There is no question that cattle ranching has left a troubling legacy in the American Southwest. This is particularly true of the corporate ranching companies of the late nineteenth century, when just two range-cattle corporations in Cochise County, Arizona—the Erie Cattle Company and Chiricahua Cattle Company—controlled well over two million acres of grazing land. Southwest rangelands have still not recovered from the overstocking of the late 1800s, making cattle ranching troublesome and unendingly quarrelsome, particularly the issue of grazing cattle on public land. Yet family ranchers are devoted, work hard, and recoil at the pejorative notion of "welfare ranching," a term grounded in a variety of subsidies, among them the so-called bargain-basement lease rates for grazing permits on public land, low property tax rates, government assistance programs, and the policy of allowing public land ranchers to use grazing permits as loan collateral. Cattle ranching has left enormous environmental costs: both habitat destruction and implications for wildlife. Environmentalists concerned about the dramatic loss of habitat and the worsening instability of what little good land remains have historically seen cattle grazing on arid lands as tragically destructive and its end long overdue. Yet the solution is not so simple.

Habitat fracturing is a serious threat to wildlife. If the working landscapes of the arid West are to remain unfractured, viable habitat, the ranching community needs to do its part and embrace the broad-scale restoration of rangeland, particularly riparian areas. Most of the Southwest's riparian habitat is on private land, and "restoration of riparian acreage is the cornerstone to the restoration of an entire watershed."[1] No other landscape feature connects ecosystems or reflects watershed conditions as effectively as waterways like streams and ciénagas and their adjacent riparian habitat. There are still ranchers who refuse to fence their cattle out of tender riparian areas, some of which are on public lease land, irking public land managers and worsening environmentalists' opinions about ranching.

The question of ranching and healthy habitat is a tangled one that has long been the subject of controversy between ranchers and environmentalists. Yet there are good reasons to support regenerative, nature-based cattle-ranching practices:

FIGURE 14 The Pitchfork Ranch has a small herd of grass-fed and grass-finished Charolais cattle, white-coated animals that have an advantage in a hotter world because they run fifteen degrees cooler than black-coated cattle. Photograph by the author.

- The sparsely peopled range avoids fragmentation of the land and allows critical wildlife corridors to remain open.
- Other than habitat-rupturing subdivisions, there are few other economic uses of the arid land of the Southwest.
- In the face of climate breakdown, there is no place for stockyard cattle in the transition to net-zero economy. Consuming less meat and dairy products is essential to achieve net-zero targets. Feeder or stockyard cattle need to become part of the past, leaving cattle ranching as the main source of beef.
- Increasing numbers of ranchers are adopting sustainable ranching practices (chapter 7) that are less taxing on habitat than historic ranching methods that damaged the Southwest.

There are a number of reasons for transitioning to range-fed and finished beef over feedlot beef:

- The cattle's food is grown on-site alongside the cattle, so the transportation-related carbon footprint is nearly eliminated.
- The Amazon rainforest and forests elsewhere are being cut down and converted to grain production to feed cattle, something that must stop if we are to avoid the massive carbon emissions and lessening of drawdown of atmospheric carbon that result from logging. The elimination of these forests will no longer have an economic driver once beef consumption declines and feedlots are eliminated.
- The cattle's higher quality life of free-range grazing leads to higher-quality beef.
- Unlike confined animal-feeding operations or factory farms, cruelty and animal-welfare deficits are absent from free-range grazing, meaning the freedom to browse in open pastures, the animal's natural living condition.
- Ranchers of range-fed cattle can ensure that they are not subjected to pesticides, hormones, antibiotics, herbicides,

fertilizers, other synthetic chemicals, or artificial supplements and foods nor fed grain.
- Hormones typically fed to feedlot cattle end up in the people who consume the beef, in turn increasing incidences of certain types of cancer, thyroid disease, obesity, diabetes, endometriosis, uterine fibroids, infertility, asthma, and allergies. Recent findings about pre- and polyfluoroalkyl substances, nicknamed "forever chemicals," will likely soon be found in stockyard beef.
- The presence of antibiotics in feedlot beef is suspected of disrupting healthy bacteria in the human gut.
- The balance of omega-6 (usually inflammatory) and omega-3 (usually anti-inflammatory) fatty acids in grass-fed beef is easier on the human digestive system.

For centuries, domestic livestock grazing has been the primary, and often only, viable economic use of rangelands, which are typically too dry for other agriculture, and is currently supporting the livelihoods of approximately a billion people worldwide—one in eight. Rangeland ecosystem services benefit several billion people globally and provide nourishment for 800 million people suffering from food insecurity. Rangelands cover more than 30 percent of terrestrial lands, including grasslands, savannas, scrub, alpine tundra, and woodlands. More than 80 percent of the world's farmland is used to feed cattle and other livestock. The industrial scale of feedlot beef production encourages Americans to eat more than their fair share of beef. The United States makes up 4 percent of the world's population, consumes 21 percent of the world's beef, uses 17 percent of the world's energy, and creates 12 percent of the world's solid waste, a weighty part of that produced by cattle.[2] Ranching will become a meaningful part of natural climate solutions only if we reduce our consumption of beef.

The solution to the cattle and carbon emissions part of the world's climate, species extinction, and soil loss crises is straightforward: eliminate stockyards and the agriculture production accompanying feeder cattle. Two dominant features of the current meat and agricultural paradigm need to be permanently edited out of our world: stockyards and disc harrowing. The clearing of forests to grow grain for cattle and soy to feed chickens and pigs must end. A 2018 study published in the journal *Science*, "Reducing Food's Environmental Impacts through Producers and Consumers," concluded that avoiding meat and dairy products is the single most prompt way for individuals to reduce their environmental footprint. Reducing beef consumption to the equivalent of two beef burgers per person per week will help reach Paris Agreement climate goals. A 2020 study by the University of Oxford and the Potsdam Institute for Climate Impact Research reported that in Western countries, beef consumption must fall by 90 percent to avoid climate collapse. The research proposed a global shift to a "flexitarian" diet; if we are to have a sustainable future, the average world citizen needs to eat 75 percent less beef, 90 percent less pork, and half the number of eggs while tripling consumption of beans and pulses and quadrupling that of nuts and seeds.[3] Our food future needs to be be far different from today's menu.

Our worsening dependence on cattle and farming might be alleviated by precision fermenting, a refined form of brewing microbes—edible microbial biomass, bacteria, yeasts, and filamentous fungi—that feed on hydrogen and methanol to create a new generation of staple foods. This process has the potential to use 1,700 times less land than the most efficient agricultural means of producing protein (U.S. soy) and up to 157,000 times less land than the least efficient (beef and lamb production).[4]

Unless or until we reach the fermenting diet, the elimination of meat—beef, pork, lamb, and chicken—and dairy products from our diets may be less daunting that it sounds, as there are an enormous variety of substitutes.[5] Cultivated meat alternatives will trigger a cascade of decarbonization throughout the globe and will slash meat production's half interest in the entire food sector's global greenhouse gas emissions. This is significant because the food sector contributes one-third of all heat-trapping gases. There's talk of a "food revolution" as millions of tons of harmless food can be produced without stockyards and factory slaughter of cattle, as well as chickenless and pigless fowl and pork. The cascade of decarbonization from meat alternatives is one of three super tipping points, along with increased electric vehicles sales and green fertilizers. There are more than a hundred cultivated ventures, suggesting the beloved market approves of this still costly and ethical food alternative.

Beyond climate and companion concerns, continuing our meat-rich diets simply makes neither economic nor health sense; meat and dairy products use 83 percent of the world's farmland and produce 60 percent of agriculture's greenhouse gas emissions while providing only 18 percent of our calories and 37 percent of protein. Between 70 and 75 percent of all U.S. beef comes from feedlots. The American brand of unregulated, consolidated, corporate capitalism has led to fewer but larger feedlots with the top three feedlot states (Texas, Nebraska, and Kansas) now marketing almost 60 percent of the cattle produced in the United States.[6] The distress of American cattle ranchers represents the underside of the staggering winnings harvested by the conglomerates that dominate the meatpacking industry—Tyson Foods and Cargill, plus a pair of companies controlled by Brazilian corporate owners, National Beef Packing Company and JBS. According to the U.S. Department of Agriculture, since the 1980s, the four largest meatpackers have exploited the governments lack of Sherman Antitrust Act enforcement, initiating a wave of mergers to increase their share of the market from 36 percent to 85 percent.[7]

A report of the Intergovernmental Platform on Biodiversity and Ecosystem Services emphasizes that the same activities driving the climate and companion crises are also worsening pandemic risk: the production and consumption of meat, palm oil, metals, lumber, and other commodities for rich countries. Scientists had long been warning that a pandemic like the one caused by COVID-19 was inevitable. Diseases like COVID-19, bird flu, and HIV are entirely driven by animals displaced from wilderness areas for cattle-food farming, logging, roads, and trade in wild species. Intrusion into wilderness has brought people into contact with dangerous microbes. Unless the destruction of the natural world is halted, an "era of pandemics" will engulf us with a vast pool of possibly 1.6 million unknown viruses, of which 600,000 to 800,000 could potentially infect humans.[8]

Long ago having committed myself to environmental concerns, I walk a tightrope with my thinking about cattle, ranching, and habitat loss. But elimination of stockyards and disc harrowing will leave beef production to ranchers where they can persevere with little impact on climate. The increased reliance on beef production in the Southwest will not only allow ranching to prosper and afford open spaces for wildlife; it will also preserve an abundance of ecosystem services. Although little recognized, ecosystem services such as erosion control, clean air, natural pollination of crops, nutrient cycling, and carbon capture, along with a host of cultural services, matter.

The Southwest range country has generally become desertified due to humankind's misuse; it's been drilled, overgrazed, excavated, bulldozed, strip-mined, overharvested, clear-cut, and polluted, practices that simply can't continue. But there's a new assailant in the neighborhood that may represent an even greater risk. This land is in danger of becoming overwhelmed by the proliferation of urban sprawl, the rash of second homes, real estate "ranchette" developments, and plop McMansions. Abusive land use includes subdivisions and starter mansions that cause habitat fragmentation and the destructive havoc that arrives when humans enter natural landscapes.

For the first time in over a century, more Americans are leaving cities to live in the country than rural people choosing urban life. The COVID-19 pandemic accelerated the transition. Americans are altering the rural world, buying small acreages on farms, ranches, and private forestland that are being irreversibly fractured. Nearly half of the 4.5 million acres of rangeland taken out of grazing in the fifteen years before 1998 were converted to urban development.[9] The U.S. Forest Service's 2000 "Rangeland Resources Trends in the United States" report found that in some scenic parts of the rural West, pastureland and rangeland values were appreciating so rapidly that ranchers were under increasing pressure to subdivide their lands for rural homes. This trend has continued and is expected to persist over the next fifty years, decreasing the amount of land available for grazing.[10] A 2021 study in *Frontiers in Forest and Global Change* found that no more than 2.9 percent of Earth's land surface is "ecologically intact" due to the sheer volume of economic activity, including urban sprawl and the invasion of rangeland.[11]

Low-density rural home development is the fastest growing form of land use in the United States since the early 1950s.[12] Once subdivisions arrive, everything changes: the wildlife habitat becomes fractured; pavement covers plant life; off-road all-terrain vehicles shatter silence and habitat; lights block the stars; forage changes; vistas disappear; watershed protection ends; corridors close; wildlife dies out; we disassociate from nature; and the metaphysical quality of open spaces vanishes. The worst of this is the destruction of the wildlife corridors locked into animals' habits of mind and on which they depend for sustaining life. The cumulative impact is devastating and irreversible. Once an unprotected farm or ranch has been sold and broken by development, that land is forever altered. The implications for biodiversity are typically disproportionately large relative to the area developed. The rapid conversion of ranchlands to rural housing developments slowed with the Great Recession, but the onslaught has returned; the ranchette lifestyle has regained favor as people escape the crowded city and the baggage that goes with it.

Let me cut to the chase: there is said to be a conflict between cattle and the part they play in the global warming crisis, but this is a flawed feud caused by the failure to distinguish range cattle from feedlot beef. It's a quarrel between committed devotees on both sides of the question. The Center for Biological Diversity and the New Mexico Cattle Growers Association see the middle ground as a sellout. But in recent decades, organizations like the Malpai Borderlands Group and the Quivira Coalition in New Mexico have staked out ground in the "radical center"—that terrain between the "cattle-free-in-93" crowd and ranchers who have dug in their heels and bowed their backs at any suggestion of change—in support of the idea that working landscapes can serve both beef production and environmental stability. Stockyards, not cattle, are the problem.

The threat of land fragmentation is exacerbated by the checkerboard nature of much of western rangeland. There are few better illustrations of this phenomenon than where we live. The Pitchfork Ranch ownership and restoration map identifies a variety of salient points: (1) private land is property settled by pioneers, who chose the most productive land nearest to water; (2) this predictable choice helped create a fractured ownership configuration; (3) isolated, small, unsettled public lands yield lower market rates; and (4) the diversity of ownership has created a checkerboarded habitat. There is patented or deeded land that Lucinda and I own, including subsurface mineral rights; federal land under the jurisdiction of the Bureau of Land Management (BLM); land owned by the state of New Mexico; and deeded land we own but for which subsurface mineral rights are either retained by the federal government or owned by someone else and subject to the General Mining Act of 1872. This is four distinct and locked-in types of landownership. Much of the Southwest is similarly checkerboarded. Land along the Burro Ciénaga watercourse is almost entirely deeded. As one moves away from water and up slope, the land becomes less productive and increasingly owned by the state or federal government. This fractured ownership template is a function of the poorly thought-out federal quarter-section laws enacted to encourage settlement of the West and a variety of strategies pursued to overcome their shortcomings. For wildlife in need of contiguous landscape and migration corridors, the additional habitat splintering introduced by further development would be a crushing blow.

Criticism of cattle ranching always circles back to the central question: "What will happen to these checkerboard rangelands if ranching is no longer profitable?" When ranchers sell out, not only is rangeland that long served as open space and provided wildlife corridors compromised, but once it's sliced up and sold to newcomers, there also follow conflicts in values with longtime local residents. Recall the Silicon Valley billionaire who purchased eighty-nine acres along a California beach and installed locked gates across private land that had given access to public beaches. There are wealthy new residents in Montana who are unwilling to have strangers use old roads for access across their property to public lands.[13] Examples are endless. These conflicts alone don't make the case for continued ranching, but with the rich overwhelming the West in pursuit of trophy ranches and large rural estates for manor homes, there are a host of long-term implications—all bad. A common perspective shared by many of the Southwest's longtime residents, who are being squeezed by developers, quoting naturalist Joseph Wood Krutch: "If people destroy something replaceable made by mankind, they are called vandals; if they destroy something irreplaceable made by God, they are called developers."[14]

Developers' mindless pursuit of real estate profits must also yield to Earth's limits, particularly in the Southwest rangeland, which is inadequate for the survival of its remaining species. Pulitzer Prize–winning scientist and writer Edward O. Wilson's 2016 book, *Half-Earth: Our Planet's Fight for Life*, proposes, "Only by committing half of the planet's surface to nature can we hope to save the immensity of life-forms that compose it."[15] The organization Nature Needs Half agrees, intending to establish itself as a worldwide movement based on the science-informed knowledge that nature needs sufficient space in order to function, that is, keeping at least half of the planet wild and intact. Although politically compromised, President Joe Biden's 2021 executive order to conserve 30 percent of U.S. land and ocean waters by 2030 reflects Wilson's thinking. Though less than Wilson's 50 percent, with only 12 percent of the country's land currently protected, the 30 by 30 goal is progress. And economic progress will follow: a report commissioned by the Campaign for Nature involving more than one hundred researchers led by University of Cambridge ecologist Anthony Waldron found that expanding areas under conservation could yield a return of at least $5 for every $1 spent and boost global economic output by about $250 million annually.[16] Those urging caution, suggesting we go slow, are those deeply invested in yesterday's sources of energy, who serve the financial interests of the privileged few and keep us on the path of destroying ourselves.

The cattle ranch is the only remaining barrier to development pressure and likely the only way to accomplish large-scale preservation of ecosystem integrity, the all-important task of restoring our dwindling open spaces in the Southwest and capturing carbon on a large scale. Ranches are open lands, archives of settlement history, biodiversity niches, watersheds, wet spots, archaeological preserves, and habitat for endangered species.[17] One of Wallace Stegner's many insights: "The worst thing that can happen to any piece of land, short of coming under the control of an unscrupulous professional developer, is to be opened to the unmanaged public."[18] Ranching counters this threat. The idea has yet to gain traction, and few, including ranchers, realize it, but ranchers are in a position to play a central role in contending with the onslaught of the climate and companion crises.

Beyond preventing fragmentation and providing a broad spectrum of ecosystem services, cattle ranching can play a role in science and education. The Pitchfork Ranch serves regional schools and universities for field trips and has hosted a scholar developing local foods, Southwest Archaeology summer workshops, regular visits by Audubon chapters, Gila River Festival tours, visits by various Native Plant Society groups, fundraising tours, and study on-site that has resulted in student reports, papers in peer-reviewed science journals, portions of several masters theses, and restoration work provided by the Aldo Leopold Charter School in Silver City. The inventories of species on the ranch maintained by ornithologists, botanists, lepidopterists, entomologists, mammologists, and herpetologists are available on the ranch website: www.pitchforkranchnm.com.

Rangeland covers 61 percent of the total land base in the eight Rocky Mountain states, the so-called dry core: Arizona, Colorado, Idaho, Montana, Nevada, New Mexico, Utah, and Wyoming. More than half of this land is privately owned, from 63 percent in Idaho to 90 percent in New Mexico. Peaking at 268 million acres in 1964, agricultural acreage declined to about 228 million acres over the next thirty-plus years. As of the end of the twentieth century, the Mountain West isn't just the fastest growing region in America; its growth rate is in the same league as those of Africa and Mexico.[19] Pressure to convert this land to housing developments and the biological importance of open spaces gave rise to the ranching and environmental interests of the Quivira Coalition and Malpai Borderlands Group, and to the pursuit of a collaborative approach to working landscapes and the environment. If habitat is to be preserved, restored, and otherwise protected for the benefit of wildlife, at-risk species, and nature generally, environmentalists need to acknowledge that ranchers typically own the best land, and if the ranchers can't make a living and are forced to sell out, fragmentation will render these lands uninhabitable for everything but people. Similarly, financial pressure on the viability of ranching is forcing ranchers to recognize there are more profitable ways to graze livestock that are better for wildlife, land values, and ranchers' reputation as land stewards. Although far from universal, the benefits of collaboration between ranch owners and environmental interests have become apparent to increasing numbers of both, drawing them closer to the radical center.

There is financial support to encourage these efforts. The National Resource Conservation Service (NRCS) and other agencies offer a variety of grant subsidies, including help to fence cattle out of riparian areas and pump, pipe, and store water to give grazers access to it without trampling watered corridors. The Pitchfork has benefited from government aid in the installation of grade-control structures that have slowed water runoff, fostering the influx of grasses and other vegetation that has improved the habitat for wildlife, as well as cattle. We have also

- Removed cattle in 2006 from the riparian area of the ranch where the ciénaga flows.
- Planted hundreds of trees and other native plants which in turn have recruited hundreds more that now line northern portions of the Burro Ciénaga.
- With financial support from NRCS, fenced cattle off two miles of the lower riverine area in 2010, allowing more than a hundred cattle-chewed stub ash trees to grow to a height of up to twenty feet.
- Planted more than a hundred cottonwood and Goodding's willow trees with installed grade-control structures and in the bosque.
- Reduced the number of cattle on the ranch from 150 cow-calf pairs to a dozen, allowing vegetation to proliferate.
- Changed the herd to all white-coated Charolais cattle to better tolerate heat.
- Ended inoculations in a program with the NRCS to increase the dung beetle population.
- With the help of BLM and NRCS, planned to add water catchments for wildlife with 3,600 gallons of storage each that will provide year-round water.

The collaborative approach to ranching and environmentalism is a new and encouraging one in this time of worsening balkanization and climate disasters. Several decades ago, if we had placed Americans on a metaphorical football field according to their political views, most people would be nestled between the two forty-yard lines; today the bulk of folks are embedded in the end zones, a nation divided by decades of misinformation, obfuscation, deception, and fraud, with people who increasingly prefer to quarrel rather than cooperate.[20] The credibility void is deep and dark, cavernous and cold: filled with disinformation, conspiracy thinking, division, cynicism, resentment, and anger. Families have stopped talking and loving one another.

Despite this polarization, seemingly disparate people working together in groups like the Quivira Coalition have allowed Lucinda, me, and others to think more positively about cattle and the potential for the future of ranching. We weren't ignorant. I have read Sharman Apt Russell's Kill the Cowboy: A *Battle of Mythology in the New West* and George Wuerthner and Mollie Matteson's *Ranching Welfare: The Subsidized Destruction of the American West*. Both books insist livestock production has torn apart the ecological fabric of arid western land. These writers maintain that half of the U.S. rangeland is degraded, with carrying capacity or ability to feed cattle reduced by over half,[21] and give overwhelming evidence that ranching has impoverished the West's biological capital. Plenty of literature documents the shortcomings of ranching in the Southwest, and ranching's destructive history can't be ignored. Opponents of ranching lament that U.S. taxpayers are subsidizing grazing-permit holders who abuse public lands.[22] Several hundred million acres of public land are being leased as of 2021, and too much of it is misused by ranchers in the West. Although cattle on public lands is a hot-button issue, it is a solvable problem too. Federal lands produce a mere 1.3 percent of the U.S. feed and forage supply for cattle and sheep. Only 2.2 percent of U.S. farms and ranches graze their livestock on federal public lands for any part of the year.[23] Despite ranchers' claims that public land is important to the future of ranching and that a vast amount of public land is burdened by livestock grazing, almost all livestock production is on private lands. The total contribution of federal public lands to the national forage, feed, and meat supply is such that the market will adjust and ranchers adapt if all public lands grazing were ended.

Changing the world's beef-raising paradigm is a challenge. The United States has more than 93 million head of cattle. Humans and their livestock now comprise almost 96 percent of all mammal biomass on Earth. The significance of this number is made clear by the same report: wild mammals account for a mere 4 percent of Earth's biomass. Born with the ability to digest almost any plant they can chew, cattle spend at least eight hours a day filling their four bellies. The 2016 greenhouse gas emissions of three meat-producing companies—Cargill, Inc., Tyson Foods, and JBS Foods—were nearly as much as any one of the biggest oil companies, like ExxonMobil, Shell, and BP. Although this figure is hotly disputed, there is an anti-cattle contingent that maintains livestock and their byproducts account for at least 32 billion tons of carbon dioxide per year. Then there is the methane, many times worse than carbon. One midsized cow or bull emits 150 kilograms of methane every year—the same environmental impact as three automobile trips from New York to Los Angeles. The planet's cows produce 150 billion gallons of methane per day; methane emissions from livestock and natural gas are nearly equal.[24]

Wuerthner, a harsh critic of ranching who argues that "the eviction of cattle is the single most important action we can take for the public lands, for the plants and wild animals," acknowledges that "the elimination of livestock grazing on public lands in the West would be of very little consequence to the overall supply of meat to the nation . . . [because] the majority of cattle are not raised here, but east of the hundredth meridian."[25] The end "of the public lands grazing system would mostly affect rich people and corporations that exploit cows as investments, as tax write-offs, or as hobbyist playthings."[26] The U.S. Department of Agriculture maintains an ongoing data base—stemming from surveys in the 16 largest cattle feeding states of about 2,000 cattle feeders with a capacity of 1,000 or more animals—of the number of cattle being fed a ration of grain, silage, hay, or protein supplements for the slaughter market.[27] As the authors of *Welfare Ranching* make clear, of those cattle, "currently, 'grass-finished' beef accounts for less than 1% of the current U.S. supply."[28] When we are confronting the part that meat and dairy play in the climate and accompanying crises, the critical takeaway is the difference between range-fed and finished cattle and stockyard feeder cattle. The bottom line: eliminating western cattle ranching would do little to reduce our global warming footprint and would catastrophically fragment the wildlife habitat and corridors of the Southwest.

CHAPTER FOUR

Species Preservation

PRESERVING WILDLIFE IS A fundamental part of the work we're doing on the ranch. In addition to improving habitat for the fauna and flora plentiful here, we are also trying to address the worsening challenge of species extinction or biodiversity loss. Although the current death toll of species worldwide is daunting and what we're doing seems small in a global context, it's something. Scientists tell us the loss of species can be arrested or slowed, but far more needs to be done, as the extinction of the world's animals is accelerating at an alarming rate. On the Pitchfork Ranch, we've had mixed results, but we've made gains in the preservation of several species, both animals and plants, that were on course to blink out but now remain, in small part, as a result of our work here.

The Aplomado falcon (fig. 15) was reintroduced on the ranch by the Peregrine Fund from 2009 to 2011. This falcon has been virtually extirpated from New Mexico, last seen near Deming in 1952. Yet it is so widespread further south that it is assessed merely as a species of least concern on the International Union for Conservation of Nature (IUCN) Red List of Threatened Species. The five-year reintroduction plan ended prematurely because these birds could not survive in the dry, climate crisis

FIGURE 15 Aplomado falcon in flight. Photograph by Alicia Arcidiacono, 2009.

habitat. We do think there may be a pair that has persisted, although we have not yet been able to photograph either bird.

Monarch caterpillars feed exclusively on the leaves of milkweed, the only host plant for this iconic butterfly species; its leaves are the only food source that supports their growing caterpillars. We had and have added more milkweed, so the monarch persists on the ranch. There are two distinct populations of monarch butterflies with different migration paths: one population (eastern, midwestern, and Intermountain West) migrates to central Mexico. The other (Pacific) migrates up and down the Pacific Coast, but not to Mexico. Because the U.S. population of the monarch butterfly has plummeted to a mere 20 percent of what it was a few decades ago, the U.S. Fish and Wildlife Service says listing the monarch as endangered or threatened under the Endangered Species Act is warranted. Yet protection has been precluded by higher priority listing actions of other government

FIGURE 16 Milkweed in flower on the Pitchfork Ranch. Photograph by Lucinda Cole, 2023.

agencies severely underfunded by politicians burdened by shallow ideology and short-term thinking.

Monarchs are important as a flagship conservation species for a variety of reasons, including pollination. They are also the only butterfly that migrates with the seasons and does so over multiple generations. They spend winter in central Mexico and fly north to the United States and southern Canada in the summer, where they breed and lay their eggs on milkweed. All species warrant preservation, but some deserve preservation if for no other reason than their inspiration and beauty.

The Chiricahua leopard frog (*Lithobates chiricahuensis*) was reintroduced on the ranch in 2006 by Randy Jennings of Western New Mexico University in Silver City. Its status under the Endangered Species Act is threatened. Every class of creature on the planet is suffering the plight of extinction, but amphibians have the unfortunate distinction of being the world's most endangered class of animal. Frogs are the most populous group in that class and have suffered the most extinctions, an estimated 170 species over a ten-year period. Another 1,900 frog species are in a threatened state, just one step below the endangered designation. Many conservation scientists worry that amphibians are facing the same fate as the dinosaurs: the extinction of an entire class of animal.

Chytrid (*Batrachochytrium dendrobatidis*, commonly referred to as *Bd*) is a fungus that is exponentially speeding the rate of a potential global mass extinction of amphibians by coating the animals' porous skin, which prevents hydration and respiration, thereby causing death by dehydration. *Bd* is responsible for the deaths of two-thirds of the amphibian species in Central and South America and has led to the decline of global populations of 500 species of amphibians in the past 50 years. The frog pictured in figure 17 is from Moreno Springs in the Mimbres River and for unknown reasons is persistently *Bd* positive. Jennings refers to these frogs living with *Bd* as "super frogs."

The Chiricahua leopard frog, native to Mexico and the states of Arizona and New Mexico, has been decimated by the chytrid

FIGURE 17 Chiricahua leopard frog. Photograph by the author.

FIGURE 18 Yerba mansa or lizard tail plant on the ranch. Photograph by Lucinda Cole, 2022.

fungus, which has led to the frog's ESA classification, in need of critical habitat. As part of a two-state conservation effort, the amphibian is now well established in the upper reach of the ranch's ciénaga. We have removed willow trees to enlarge open areas along the banks of the ciénaga, providing warm places for the frog to rest and sun.

Although the yerba mansa plant (*Anemopsis californica*) is not at risk, it is uncommon, and has medicinal purposes. It was introduced on the ranch by Coyote Phoenix in 2012 and continues to flourish. Uses of the plant vary: antibacterial, antimicrobial, and as a treatment for a variety of inflammatory ailments.

Listed by the IUCN as endangered, Ray Turner's spurge plant (*Euphorbia rayturneri*) was previously unknown to science when discovered on the ranch in 2009. For several years, until it was located at two other sites in New Mexico, the Pitchfork Ranch was the only place on the planet where it was known to exist. This new spurge was discovered during one of the Gila Plant Society's periodic visits to the ranch. The society went through a challenging effort to acquire formal recognition of the plant

FIGURE 19 Ray Turner's spurge plant, discovered on the ranch. Photograph by the author, 2009.

as a new species because there initially was skepticism among experts that it was in fact new. Initial assessments considered it merely a sample of a known spurge, but because of persistence by Gene Jercinovic, a member of the crew who found it, the plant was eventually recognized as a new species.

Raymond M. Turner was a plant ecologist with the U.S. Geological Service until his retirement and is the author or co-author of a number of books. *The Changing Mile* (1964), *The Changing Mile Revisited* (2003), and *The Ribbon of Green: Change in Riparian Vegetation in the Southwest United States* (2007) are the most important to this work, addressing the landscape change that mistakenly seems static within a single lifetime. *The Changing Mile* juxtaposed photographs of the Sonoran Desert region in the late 1880s and early 1900s with then-current same-location photographs, demonstrating the extent of change caused by human misuse and climate shifts. The *Revisited* book added photographs from the early aughts, three decades later, along with recent information about plant ecology, land use, and climate to help understand the causes of the changes that have occurred. The third book, *The Ribbon of Green*, similarly uses same-location photography to document changes in riparian reaches in the Southwest, such as the waterway on the Pitchfork Ranch. In chapter 6, you'll see how we utilize this technique to demonstrate improved habitat as a result of restoration.

Wright's marsh thistle (*Cirsium wrightii*) was reintroduced on the ranch in 2009 by New Mexico botanist Bob Sivinski and Phil Toone. At the behest of WildEarth Guardians, the U.S. Fish and Wildlife Service proposed to list the thistle as threatened under the ESA. It is known to exist in only eight locations in New Mexico and appears to have blinked out here—possibly due to fast flood flows that persisted in the early phase of restoration.

The Gila topminnow (*Poeciliopsis occidentalis*, fig. 21), at one time the most abundant fish species in the Gila River basin, was reintroduced on the Pitchfork Ranch in 2008 by Yvette Paros, then of the U.S. Fish and Wildlife Service. For a number of years, the ranch was the only location where the fish could be found

FIGURE 20 Wright's marsh thistle. Photograph by Lucinda Cole, 2012.

FIGURE 21 Gila topminnow. Photograph by Lucinda Cole, 2015.

in New Mexico. It has been listed as endangered under the ESA since 1967. Long extirpated from New Mexico, relocated to Arizona, and now returned to its native habitat, the fish was well reestablished here. On several occasions, we relocated many hundreds from locations that were drying out after flooding deposited them down-channel, but it appears that they too may have blinked out. Another reintroduction effort is in the works.

The viceroy butterfly (fig. 22), which, though not closely related, gives the impression of a sibling of the monarch, queen, and soldier butterflies, was on the ranch when we arrived. It is a species of concern in New Mexico. They remain abundant on the

FIGURE 22 Viceroy butterfly. Photograph by Lucinda Cole, 2006.

north portion of the ranch and are often seen at the headquarters and farther south. We initiated a reintroduction effort, but an expert soon found that the Viceroy was already here.

Referring to the endangered Owens pupfish, Christopher Norment, an ornithologist who has studied desert fishes in Death Valley, California, and Nevada, wrote, "To protect the fish is an act of hope, a sacrament and celebration."[1] To restore habitat is also an act of hope, a commitment to preserve life and to assure our grandchildren a safe and healthy place to live and raise their own children. In the face of fierce pressure and insurmountable odds, the Chiricahua leopard frog—pictured in the wood engraving at the opening of this chapter—and several other species on the Pitchfork Ranch are persisting, surviving, and promising that life on the planet has a habitable future.

CHAPTER FIVE

The Trifecta Crisis

THE RELEASE OF METHANE, increased carbon dioxide, and other gases in the atmosphere reduces the amount of solar energy Earth radiates back into space. This is the greenhouse effect that produces a hotter Earth. A warmer atmosphere holds more vapor. There is more water moving around, so the water fluctuations throughout the globe get bigger; more rain and wind cause increasing severe weather disruptions and meteorological extremes. Today's atmosphere contains about 5 percent more moisture than it did a century ago due to warming. With this increase in temperature and more moisture in the atmosphere, the planet has experienced a *doubling of the rate* of record-breaking heat over the past several decades. Earth's water cycles intensify as the planet warms and boosts Hadley's trade winds and other weather currents like El Niño and La Niña. More extreme and disruptive weather is triggered: floods, fires, heat waves, mudslides, volcanic eruptions, earthquakes, beetle-driven forest die-offs, tornadoes, hurricanes, and storms of both dust and rain that are increasingly frequent and severe. The climate crisis is so dire that even if we immediately did everything scientists warn is necessary, these extreme weather events will get worse.

With the onslaught of ruinous weather in the second half of 2021 and throughout 2022 and 2023, little needs to be said about the existence or severity of the climate crisis. Most thoughtful people agree with this appeal from more than eleven thousand scientists in the Alliance of World Scientists:

> We declare . . . clearly and unequivocally that planet Earth is facing a climate emergency. . . . The climate crisis has arrived and is accelerating faster than most scientists expected. . . . It is more severe than anticipated, threatening natural ecosystems and the fate of humanity . . . [causing] significant disruptions to ecosystems, society, and economics . . . potentially making large areas of earth uninhabitable. . . . The crisis is closely linked to excessive consumption of the wealthy lifestyle. . . . An immense increase of scale in endeavors to

> conserve our biosphere is needed to avoid untold suffering due to the climate crisis. To secure a sustainable future, we must change how we live . . . [transform] the ways our global society functions and interacts with natural ecosystems . . . [and address] increases in both human and ruminant livestock populations, per capita meat production, world gross domestic product, global tree cover, fossil fuel consumption, the number of air passengers carried, carbon dioxide emissions, and per capita CO_2 emissions.[1]

The sixth report of the 2021 Intergovernmental Panel on Climate Change makes clear the climate is overheating faster than earlier reported, emphasizing that every ounce of emissions matters. United Nations Secretary General António Guterres was blunt: "This report must sound a death knell for coal and fossil fuels, before they destroy our planet. . . . [It's] a code red for humanity. The alarm bells are deafening, and the evidence is irrefutable: greenhouse gas emissions from fossil fuel burning and deforestation are choking our planet and putting billions of people at immediate risk."[2]

John Kerry, U.S. President Joe Biden's special envoy for climate, said, "The IPCC report underscores the overwhelming urgency of the moment."[3] There is no longer any question: the unprecedented human-caused carbon dioxide levels are higher than they've been in at least 2 million years, temperatures are higher than they've been in at least 100,000 years, and the sea level is rising faster than it has in at least 3,000 years. We're stumbling over the precipice, threatening sustained, organized human life.

The danger posed by species extinction, soil loss, and soil depletion is less understood than the climate crisis, but the three are interrelated. It's instructive to think of these three crises as a trifecta of risk. A trifecta is a wager in which a person picks the first three finishers in a race in the correct order. In this instance, the order of finish doesn't matter; we only have until 2030 to adopt radical and lifesaving changes. This is a bet we are losing, in part, because of deceptive language. Despite the heat, fire, and a water scarcity, we're not in a drought. The so-called mega-drought occurring across the Southwest, now in its twenty-third year, is the first in more than five hundred years, and may be the worst in human history. While it's common to hear the term *drought* or *extended drought* used to describe the overheated climate, a drought is a period of abnormally low rainfall leading to a water shortage, and a *period* is a set length of time. A period is what you find at the end of the first quarter of a basketball game or the end of a sentence. Climate scientists are not warning us about a specific period of extreme weather; we are not living through an interval or era with an end. What we've seen for at least the last decade will continue. This is the ever-worsening lethal weather that is best thought of as the Trifecta Crisis.

THE CLIMATE CRISIS

Cofounder and president emeritus of the Land Institute Wes Jackson has coined the most concise judgment of our plight: "We live in the most important moment in human history."[4] The entire biosphere—the frail membrane surrounding Earth in which life occurs—is in jeopardy. Scientist and policy analyst Vaclav Smil sees these crises plainly: "Without a biosphere in a good shape, there is no life on the planet. It's very simple. That's all you need to know."[5] We're the sole cause for myriad threats, leading biologist, ecologist, and naturalist Edward O. Wilson to contend we are "the most destructive species in the history of life."[6] We're living in a carbon-dioxide-saturated world, in the most perilous moment ever; the era of high-carbon lifestyles is exhausting the potential for life on the planet. Why a democratically elected government and its agencies foster the unimaginable profits horded by the fossil fuel industry and the end of civilization, as we know it, is a question beyond the scope of reason. The combustion age must end.

David Wallace-Wells begins his 2019 book, *The Uninhabitable Earth: Life after Warming* with this: "It is worse, much worse, than you think."[7] He goes on to describe the unthinkable consequences of the climate crisis and the need for immediate and radical action to avoid exposing the planet to unimaginable destruction. We are bound together in a mass of climatological and geological agents—humans have ganged up on life—changing the most basic physical processes of the planet. Unregulated "fossil capitalism" has led to a population of climate vandals that has broken the planet. Our failure to take charge of these crises suggests the ongoing collapse of our social and cultural order.

In a favorable turnabout, the more recent "Beyond Catastrophe: A New Climate Reality Is Coming into View," Wallace-Wells wrote that earlier climate projections looked apocalyptic and our "business as usual" way of life was leading to civilizational collapse. In a late 2022 *New York Times Magazine* article, he wrote that astonishing declines in the price of renewals, global political mobilization, and adjustments to the input assumptions of energy models has created a less destructive picture and thus is increasing cause for hope. This is welcome news, yet, absent an abrupt and decisive about-face, crisis looms.

British naturalist Sir David Attenborough told the twenty-fourth Conference of the Parties to the United Nations Framework Convention on Climate Change (COP24) in 2018 that we are undergoing a "disaster of global scale, our greatest threat in thousands of years" and if there is not a complete transformation of the global economy, we are headed for a "collapse of our civilization and the extinction of much of the natural world."[8] This unfolding atrocity has no geological analog—no volcanic eruption or asteroid—that caused earlier extinctions. As baffling as this thought is, we humans are exterminating ourselves. This can't be said enough: survival requires that we rejoin the community of life on this planet and abandon the illusion of humankind as overlord. Care and repair are the orders of this fragile moment.

Far worse than today's gun violence in schools, the January 6 insurrection, the Ukraine War, or the COVID-19 pandemic, what troubles me deeply is that we are hooked—absolutely, resolutely, adamantly—on *convenience.* Our unwillingness to change, to be inconvenienced, concerns me more than any other impediment to solving the climate and companion crises. We've become saturated in chronic convenience, analogous to an addiction against which reason is powerless. I'm not sure we can save ourselves. Too few have shown a willingness to carpool, eat less meat, forgo airline travel, abandon plastic and single-use products, or consume less. Difficult and unending, these are habits we need to break and the changes we have to make.

In *The Dawn of Everything*, David Graeber and David Wengrow write: "One of the things that sets us apart from non-human animals is that animals produce only exactly what they need; humans invariably produce more. We are creatures of excess."[9] We've not been able to trigger the reptilian fight or flight response these crises clearly call for. This is where unregulated capitalism, consumption, and convenience bear on the Trifecta Crisis calculation: "excess" as the exclusive explanation for greenhouse gas emissions. You probably wouldn't be reading this book if you weren't aware that we're in a serious predicament. If you're wondering how we got here and why we are stuck here, I hope the answer will prompt you to do more about solving these problems than you may be doing now.

When considering how best to respond to these three environmental crises, it helps to keep in mind that this age of humankind's overheating the planet is a tangled, structural *multiplex* of factors that has left us with a cascade of previously unrecognized consequences. These surface with frightening regularity: warming temperatures that put people at risk of giving birth up to two weeks earlier; underweight or stillborn babies; altered animal migration; redistribution of weight on the planet; flaking and disappearance of the world's oldest cave paintings; birds having 25 percent fewer chicks and laying eggs twelve days earlier; some birds getting smaller, others larger; increasing shark attacks; the crumbling of the Swiss Alps; warmer sand during the incubation period of loggerhead turtles in Cape Verde leading to 84 percent being born female; wheat and corn turning toxic as a defense mechanism against extreme heat; frogs croaking at a higher pitch; fungal pathogens moving to higher latitudes at a rate of about seven kilometers a year; warmer air increasing baseball homeruns; and possibly (suspected, yet unproven) even an increase in a rare brain-eating amoeba killing humans and rare microorganisms known as golden algae causing the die-off of tons of fish and mollusks. Arguably worst of all, ancient bacteria and viruses released in melting ice could devastate modern humans, who have no natural immunity to diseases common hundreds of thousands of years ago.

Svitlana Krakovska, Ukraine's leading climate scientist, "started to think about the parallels between climate change and this war and [said] it's clear that the roots of both these threats to humanity are found in fossil fuels. . . . Burning oil, gas and coal is causing warming and impacts we need to adapt to. And Russia sells these resources and uses the money to buy weapons. Other countries are dependent upon these fossil fuels, they don't make themselves free of them. This is a fossil fuel war. It's clear we cannot continue to live this way; it will destroy our civilization."[10]

Woefully under-recognized, the complex of causes for these crises is bred into the social and economic fabric of growth, progress, excess consumption, and the conveniences of modern life. The phrase "mobility-related consumption" helps capture the fact that the lifestyle of the well-heeled—best illustrated by air travel—is the main cause of these crises. Most people in the world have never set foot on an airplane. Nearly 90 percent of the global population rarely flies: In 2018 only 11 percent took a flight. Only about 3 percent of the world's population travels internationally by plane. Air travel accounts for 4 percent of greenhouse gas emissions, yet a mere 1 percent of the world's population accounts for half the flight-caused greenhouse emissions.[11] These select few, many owning multiple homes and traveling by private jet, are the most obvious example of the "super-emitters" most responsible for these crises. Private jet sales are on track in 2023 to reach their highest level ever, forecast to reach almost $35 billion. The size of the global fleet has more than doubled in the last two decades. Wealthy jetsetters contribute 225,000 times as much global warming as one of the world's poor. Elon Musk has a new jet and made 171 flights in 2022, including one that lasted only 6 minutes and was responsible for more than 2,112 tons of carbon dioxide emissions, 132 times more than the entire carbon footprint of an average U.S. resident.[12]

If you're reading this book, you likely fall into the world's top 1 percent of income earners—those of us who earn more than about $34,000 a year. Cinda and my yearly Social Security income exceeds that figure. There is debate about this $34,000 yearly income figure as the cutoff for the world's top 1 percent, most of it coming from those unwilling to make the changes necessary to curb these crises, but the essential point is that most Americans make the most money and thus make these crises worse than most people. The 2020 Oxfam report reveals that One Percenters not only have more wealth than the bottom 99 percent but were responsible for more than twice the greenhouse gas emissions than the poorest half of the world's population generated from 1990 to 2015. The richest 10 percent of the global population, comprising about 630 million people, were responsible for about 52 percent of global emissions over a recent twenty-five-year period, while the poorest 10 percent cause about 5 percent. The late 2022 Oxfam report on the investments of 125 of the world's richest billionaires shows they are responsible for emitting 393 million metric tons of carbon dioxide a year, more than a million times the average of those in the bottom 90 percent of the world's population.

A closer look at the wealth and income disparity reveals even more troubling implications. The phrase "We Are the 99

Percent" gained currency after the bailouts following the 2008 financial crisis, capturing the anger and anxiety felt by those who understood the masses were being shortchanged in the rising inequality of an unfair economic system—and nobody went to jail. Yet the phrase obscures the distinctions between the 1 percent, 0.1 percent, 0.01 percent, and 0.001 percent. A detailed slicing of the wealth of these favored few is beyond our needs, so let these facts suffice: (1) there are 788 American billionaires and 2,755 billionaires worldwide; (2) a new billionaire is minted every other day; (3) the wealth of the .01 percent has ballooned by a staggering $1.8 trillion since the start of the COVID-19 pandemic to 2023; (4) the 10 richest men doubled their fortunes during this period while the incomes of 99 percent of humanity fell; (5) the 25 wealthiest American enjoyed a 3.4 percent tax rate; (6) a 2022 study of 300 top U.S. companies found that the average ratio of CEO pay to median worker pay jumped to 670 to 1, up from 604 to 1 in 2020, and 49 of those firms had ratios above 1,000 to 1; (7) in 2017, 44 individuals inherited more than $1 billion each, totaling $189 billion; and (8) entry into the upper 10 percent—requiring a minimum of $1.2 million in assets as of 2021—now takes 24 times the wealth of the median household, while entry to this group took 10 times the median household wealth in 1963. The implications of these numbers become clear with recognition that a billion is a thousand million.

The point: access to the life of the world's "better off" has become illusory, and admission to the top 10 percent has become almost impossible for the majority of the world's population. The widespread implications of this "new aristocracy" are at the center of the crises facing life on the planet. The myth of merit, failure to recognize privilege, and abandonment of social justice increasingly isolate the deciders and polluters from everyone else. White people are eight times more likely to belong to the affluent few than are people of color. The cost-of-living crisis has worsened for multitudes in the United States, a country where increasing numbers of people are burdened with choices exclusively about survival, where the possibility of lives of fullness, meaning, or imagination is receding over time. Those of us not suffering the heat and consequences of these crises and who fail to see our inordinate share of responsibility have shown little interest in changing.

The manifest amorality of this wealth disparity, the extreme reach of the One Percent, manifests in a variety of ways. The One Percenters possess half of all stock owned by Americans. In 1982 less than 10 percent of campaign funds came from the top 1 percent, a number that jumped to 46 percent by 2018, with 22 percent coming from four hundred mega-donors. The January 2022 Oxfam study reports that extreme inequality contributes to the death of at least 21,000 people each day, or one person every four seconds. Who among us feels responsible for that? At Yale and Princeton, more students come from the top 1 percent of income families than from the bottom 60 percent. Two-thirds of students in Ivy League schools come from families in the top 20 percent. One study found that sons of senators have an 8,500 times higher chance of becoming a senator than an average American male. American sons of chief executives are 1,895 times more likely to become one; sons of governors, 6,000 times more likely to be a governor; sons of generals, 4,582 times more likely to become an army general; and sons of billionaires, 28,000 times higher chance of being a billionaire. These numbers put flesh on the ideas that merit is a myth and that our problems are structural and deep-rooted.

One of the most common ways to avoid accepting the unassailable fact that the climate is in crisis is to suggest overpopulation is the problem, but those making this claim are typically among the world's thin slice of $34,000-a-year income earners who are the primary cause of these problems. This group hasn't faced up to the truth that it is the source of the problem and is failing to take responsibility to abandon the convenience of its overly consumptive lifestyle, an inherited way of life to which we've all grown accustomed. The 90 percent of the world's population stuck at home, too poor to travel, are not the problem;

it's the "mobility-related consumer" who is causing these crises. A 2020 study in the journal *Nature Communications*, "Scientists' Warning on Affluence," confirmed the alarm of those eleven thousand scientists quoted at the opening of this chapter: they concluded—and this is the most important sentence in the book—"*by far the most fundamental driver of environmental destruction is the excessive consumption by the wealthy.*"[13]

While that assessment emphasizes the jet set, the consumption by us $34,000-plus earners is among the chief drivers of industrial agriculture, stockyard beef, deforestation, fossil fuel use, and other major sources of environmental degradation. Capitalism has fostered a global structural inequality where the richest people have caused the Trifecta Crisis. The rich are *combusting* others to death. This imbalance is intimately related to the absence of environmental stability that is threatening the very existence of human societies. The *Nature Communications* paper zeroes in on the atrocity of affluence and takes the favored few to task:

> It is clear that prevailing capitalist, growth-driven economic systems have not only increased affluence since World War II, but have led to enormous increases in inequality, financial instability, resource consumption and environmental pressures on vital earth support systems. . . . Consumption is more aptly labelled as affluence. . . . Avoiding consumption means not consuming certain goods and services, from living space (overly large homes, secondary homes of the wealthy) to oversized vehicles, environmentally damaging and wasteful foods, leisure patterns and work patterns involving driving and flying. . . . Long-term and concurrent human and planetary wellbeing will not be achieved in the Anthropocene if affluent over-consumption continues, spurred by economic systems that exploit nature and humans.[14]

There is a difference between *luxury* and *substance* emissions, between wants and needs.

The promise of neoliberalism was that trust in the market gave everybody a fair shake. The reality is that oligarchic wealth, inequality, bitter class war, and poverty have brought us to this moment's slender window of time when humans can still alter the planet's downward trajectory. When we come to grips with our part in these crises, how long this threat has been known, who lied about it or did nothing about it, then changing our consumptive, polluting behavior and taking to the streets will be easier. If we don't face up to this inherited diet of convenience and jettison the habits it fosters, we'll not secure the survival of our children's children. To save them, the transition out of our carbon world requires an informed, imaginative leap into a safer lifeway of modesty, thrift, and inconvenience.

You and I and our families are likely among the 10 percent who are causing half of the greenhouse gas emissions, who fly the most, and who thus have a far greater moral obligation than most to right this listing ship. Having the wherewithal to acquire this ranch puts Lucinda and me in the group responsible for causing so much damage and doing so little about it. Before retiring, we felt little disquiet about the state of the world as we unwittingly contributed to the climate and companion crises. Our time here has taught us how we can do more. We've stopped flying, cut back beef consumption, recycle and volunteer at recycling drop-off, and have adopted a number of practices like avoiding single-use products, purchasing fewer clothes, and mending the ones we have. We hope to capture our lifetime of carbon dioxide emissions before life's end.

Understanding carbon and capital—the wrong of economic growth and the convenience of consumption—is the key to understanding these crises and their cure. Strange but true, in the face of these carbon-caused crises, carbon is what makes up life as we know it. We are made of carbon. Life exists on Earth only because the carbon atom possesses certain exceptional qualities. It's simple: no carbon, no life.[15] But it's the link between carbon and capital that is set to sink the ship. Just about everything we've made in the past two centuries, we've made from fossil fuels, leaving atmospheric carbon in our wake: energy, cars, equipment, food, medicine, plastic, and consumer goods. Are we really capable of making structural changes in the system that has provided us with so much? If we persist in our "business as usual" lifeway, certainly not.

Corporations like ExxonMobil funnel huge amounts of money to have their way while pursuing a greenwashing approach to the Trifecta Crisis and continuing to drill and spew. A recent podcast with research entomologist and agroecologist Jonathan Lundgren addressed regenerative agriculture and how and why it's superior to industrial agriculture: "I'm a scientist . . . so data is really important and it is absolutely necessary, but it is not sufficient to change a human heart. . . . and mind . . . and that is ultimately what we have to do to correct the trajectory that we're on as a planet."[16] Of course, he's right, but long ago, our Supreme Court foolishly created the fiction in the world of U.S. capitalism that corporations are people. Our dilemma is that corporations don't have hearts.

If we're going to overcome our collective irresponsiveness to these crises—the mass delusion of normalcy—it's helpful to know why we've suffered such fatal delay in recognizing them. Corporate-owned, fossil-fuel-entrenched mainstream media didn't merely poison the well with caution and disinformation but emptied it of meaningful knowledge. While arguably too late, a round of applause is due the *New York Times* and Nathaniel Rich for publishing the August 2018 article "Losing Earth: The Decade We Almost Stopped Climate Change"—now a book, *Losing Earth: A Recent History*. It's a full-throated review of the history of climate science and political debates that went on from 1979 to 1989. This was a time when we knew the basics of what we know now and failed to act. Rich details the enormous amount of scientific effort that went into understanding the risks greenhouse gas emissions pose to the planet and the failed efforts to incorporate that science into policy. Everyone at the table agreed on the facts, but it's said they could not agree on a single sentence to describe the way forward. It's a pathetic tale of the unscrupulous crisis-denial community and the oil-drenched Bush-Cheney White House.

James Hansen explained this to the U.S. Senate in 1988, yet like John Wesley Powell before him, he was ignored. Hansen and sixteen other climate scientists released a paper in 2015 projecting sea levels will rise as much as ten feet in the next fifty years. Another paper by Hansen and eleven experts—"Young People's Burden: Requirement of Negative CO_2 Emissions," prepared for use in the Oregon youths' climate crisis lawsuit against the federal government—states the 2016 average temperature is likely the warmest since the Eemian Period, an interglacial era ending 115,000 years ago, when there was less ice and the sea level was 20 to 30 feet higher than today. These studies warn that current greenhouse gas reductions, which are supposed to limit average warming to the internationally agreed-upon standard of 2 degrees Celsius, possibly 1.5, are not strong enough. A rise of just 1 degree Celsius could trigger sea-level increases as high as 16 to 30 feet. America's most recognizable name in climate science, James Hansen, maintains, "The message for policymakers is that we have a global crisis that calls for international cooperation to reduce emissions as rapidly as practical."[17] If sea levels rise as much as 5 or 6 feet, dozens of world cities—New York, Tokyo, Hong Kong, Cape Town, Bangkok, and Alexandria, among numerous others—will be swallowed by the sea in our children's lifetimes.

Not a single science organization in the world disagrees with the consensus view that the climate crisis is real, dangerous, and human caused. All national academies of science and meteorological associations of every country in the world hold the view that the climate is warming and that this warming is caused by human activity. The list of organizations is nearly two hundred strong, including these U.S. organizations: the Department of Energy, the Department of Defense, the Navy, the National Science Foundation, the National Oceanic Atmospheric Administration, the National Aeronautics and Space Association (NASA), and the National Weather Service.

SPECIES EXTINCTION: THE BIODIVERSITY CRISIS

The 2022 Living Planet Report of 32,000 populations of 5,230 animal species reported Earth's wildlife populations have declined 69 percent in the last half century, up from 60 percent in 2018, while the Caribbean and Latin America region has suffered a 94 percent decline in 48 years. Among the most severe examples, in Alaska, the fastest-warming state in the United States, the Arctic snow crab population—actually four species, Bearing Sea snow crab, Bristol Bay red king crab, St. Matthew Island blue king crab, and the Pribilof Island red and blue king crab in the eastern Bering Sea—plummeted 90 percent between 2019 and 2021. This animal is said to be a canary in the coal mine for cold water species. This is being called the Mass Extinction Crisis.

A 2019 report released by the United Nations found one million land and marine species across the globe are threatened with extinction. Land and sea animals worldwide are roughly 50 percent less abundant than forty years ago. Freshwater fish are in decline; almost a third of global fish populations are in danger of extinction; large fish populations are down 94 percent; and sixteen freshwater fish species were declared extinct in 2020. Eighty-six percent of North American bird species are at risk. The Intergovernmental Platform on Biodiversity and Ecosystem Services 2020 report assessed changes over the past five decades and found nature declining globally at rates unprecedented in human history. The Convention on Biological Diversity issues periodic reports summarizing data on the status and trends of biodiversity. Its 2020 report, "Global Biodiversity Outlook 5: Humanity at a Crossroads," concludes the world failed to meet any of the established targets to stem the destruction of wildlife and life-sustaining ecosystems in the last decade. About 20 percent of the world's countries are at risk of ecosystem collapse. There's talk of an insect apocalypse. The 2021 Chatham House thinktank report, supported by the United Nations, concluded the main threat to 86 percent of species known to be at risk of extinction is agriculture.

Elizabeth Kolbert's 2014 Pulitzer Prize–winning *The Sixth Extinction: An Unnatural History* maintains that 20 to 50 percent of all living species on Earth will be extinct by the end of the twenty-first century. As Kolbert's title suggests, the current loss of species is being referred to as "the sixth extinction crisis"—following five known major extinction waves in geological history—and the numbers are terrifying. The background rate of species loss is the extinction pace before humans arrived on the planet. Focusing on the death rate of extinction rather than the number of species lost, it is now thought species are disappearing far faster than biologists believed, adding intensity to the view that the world is in the throes of a sixth great extinction and that scientists have been overcautious. A recent report published by the IUCN Red List warns that although extinction is a natural process, "the rapid loss of species we are seeing today is estimated by experts to be between 1000 and 10,000 times higher than the 'background' or expected natural extinction rate. . . . Unlike the mass extinction events of geological history, the current extinction phenomenon is one for which a single species—ours—appears to be almost wholly responsible."[18] Accelerated extinction means one million land and marine species could become extinct due to human actions unless transformative change is made across local, national, and global levels.

A review of scientific literature by an international team of scientists titled "Defaunation in the Anthropocene" focuses not on *rate* of extinction but on *numbers* of species that have died, lending further support for the threat of a massive die-off. They too propose Earth is in the early stages of a species extinction: "We live amid a global wave of anthropogenically driven biodiversity loss [in which] human impacts on animal biodiversity are an under-recognized form of global environmental change."[19] *Nature* published a study warning the sixth great extinction is looming, with a staggering 41 percent of all amphibians on the planet now facing extinction while 26 percent of mammal species and 13 percent of birds are similarly threatened. The loss of more than half the planet's wildlife in less than fifty years and a future of ongoing comparable estimates is catastrophic. One example of this accelerated rate and associated frightening numbers: an estimated one billion animals perished in the 2019 to 2020 Australian bush fires. We're in the midst of a historic, chronic planetary emergency—an illness of disruptive climate heat, pathogens, habitat disappearance, species extinction, and soil loss—that requires a warlike mobilization.

American editor and writer Norman Cousins's Saturday *Review* editorial of nearly forty years ago recognized, "History is a vast early warning system."[20] John Adams saw history as "a lamp of experience."[21] Even John Wayne understood this: "Tomorrow hopes we have learned something from yesterday."[22] History is the compass we can use to negotiate the perils of the present. Absent historical literacy, it's inevitable the past will plague us. History used to record how far we had come from barbarism—a matter of pride, progress, and hope—but now history warns us just how fragile civilization can be. It's currently a matter of anger, guilt, and shame as we come to terms with the extreme climate cruelty of the fossil-fuel industry and its co-conspirators. The historical lesson about greenhouse gases is that they have been the cause of so many past species extinctions, although those were far less accelerated than the current set. The warning lamp of history has been ignored before and often, but humankind has never turned a blind eye to such consequential danger as the climate and companion crises we are facing now.

Cristiana Paşca Palmer, executive secretary of the United Nations Convention on Biological Diversity—the world body responsible for monitoring the planet's biodiversity—tried to warn us we had only two years, until 2020, to secure an agreement to stop the "silent killer" of biodiversity collapse that she maintains is as dangerous as climate breakdown. We failed. The loss of biodiversity has humankind on a trajectory to be the first species to document its own extinction. Conservationists had long been desperate for a last-ditch biodiversity accord similar to the 2015 Paris Agreement to thwart dangerously high rates of animal and plant loss. Intimately interconnected with nature, we are dismantling the infrastructure of life. Finally, the 2022 Kunming-Montreal Global Biodiversity Framework, signed by 140 countries, was a firm first step in resetting our relationship with the natural world, a landmark agreement to protect 30 percent of the planet's lands, costal lands, coastal areas, and inland waters by 2030.

Why does it matter when species are removed from an ecosystem? The clearest answer mirrors molecular biologists' discovery that everything about human life is regulated at the molecular level. As Sean B. Carroll points out in *The Serengeti Rules: The Quest to Discover How Life Works and Why It Matters*, "just as there are molecular rules that regulate the numbers of different kinds of molecules and cells in the body, there are ecological rules that regulate the numbers and kinds of animals and plants in a given place." It turns out that the physiological rules or regulations for our bodies are surprisingly analogous to the ecological rules running the planet. There is an inherent wisdom in these commands or rules, and this "logic explains, for example, how our cells or bodies 'know' to increase or decrease the production of some substance. The same logic explains why a population of elephants on a savanna is increasing or decreasing."[23]

Just as the absence of a regulatory mechanism in a cell or the presence of something foreign can cause cancer, the absence of a regulatory or keystone species can have similar implications for the environment. This explains why the absence of the wolf in Yellowstone National Park led to the decline in willow trees and why their reintroduction initiated the reduction of elk to previous numbers, thereby decreasing the elks' browsing on trees, fostering the return of the willow. Elk and willows play a critical role in wolves' success in the Yellowstone ecosystem: willows serve as browse for elk, and elk serve as food for wolves. Scientists initially thought the return of the wolf alone would lead to a reduction of elk: fewer elk and less willow browsing would restore the ecosystem. Despite improvement, it was found the return of the wolf was not enough to completely restore willows because willows needed sluggish streams for successful reproduction. This required the presence of beavers. But like wolves, beavers were gone too. Once beavers returned and reinstalled their dams, stream waters slowed, seeds dropped, the system regulated, and the ecosystem's health continued its return.

We are having a similar experience on the Pitchfork Ranch. We can't have beavers because of a neighbor's opposition—consent of landowners within a five-mile radius is required under New Mexico law—but installation of grade-control structures in the incised ciénaga serves a similar purpose. Now regulated, the

FIGURE 23 A perpendicular post vane grade-control structure installed at around mile 5 of the 8.3-mile reach of the Burro Ciénaga on the ranch. Photograph by Lucinda Cole.

water flow has slowed, wicking has increased, soil has aggraded, seeds have dropped, and vegetation has spiked as the ecosystem has begun to restore itself (fig. 23). This grade-control structure consists of twenty-five six-foot-long juniper posts installed in a backhoe-dug trench, with eighteen inches of the posts remaining above ground to slow flood water. The water reaches the structure, slows, and drops rabbitbrush or chamisa seeds, resulting in the plush plant growth above the structure with few plants below. Like the grade-control structure, new growth also captures debris and sediment, further slowing future flows, causing further soil aggradation, moisture capture, and plant growth.

Sean B. Carroll's *The Serengeti Rules*, describing the rules of regulation, offers a number of examples illustrating the discovery of dysfunction in both cells and habitats. Skipping examples in the cellular world and focusing on plants and animals:

- Removal of a predatory starfish in a twenty-five-foot-long by six-foot-deep area at Mukkaw Bay of the Olympic Peninsula, Washington, reduced the diversity of the intertidal community from fifteen to eight species. While the population of mussels—the starfish's natural food—increased rapidly, one predatory snail increased ten- to twenty-fold. Absent starfish, the mussels eventually eliminated all other species.
- Sea otters eat urchins, which eat kelp. Without otters, urchins overwhelm the habitat, and kelp become absent. In the Aleutian Islands, if there are no otters, there is no kelp; and if there's no kelp, there are no bald eagles.
- In a freshwater stream in Oklahoma, predatory, herbivorous algae regulate the abundance of minnows and plants.
- Wolves positively influence fir tree growth in Michigan by controlling the density of moose.

The results of these experiments show how one predator can regulate the makeup of species by what it eats and doesn't eat. Ecosystems collapse without predators. These studies also force us to recognize the implications of our apathy, carelessness, and indifference to the current species loss emergency: "For most of the twentieth century and across much of the planet, humans have hunted, fished, farmed, forested, and burned whatever and settled wherever they pleased, with no or very little understanding or consideration of the side effects of altering the populations of various species or disturbing their habitats."[24]

Those who study complex, self-regulating systems such as atmosphere, ocean currents, soil, and ecosystems have made clear that our lives are entirely dependent on these systems. Operating according to certain mathematical rules, these systems regulate themselves under normal conditions until exploited beyond balance and tipped into a new state of equilibrium which is destructive and often irreversible. What complicates our understanding of these regulatory mechanisms is that it's not just the function of a single species that rules. It is not merely the absence of the wolf that sickens the Yellowstone habitat; rather it's the wolf and the beaver and additional species that serve to regulate the habitat. Carroll offers a deeper explanation for why all forms of life are important: "*It turns out that life—from the molecular scale all the way up to the ecological scale—is usually governed by longer chains of interactions than we first imagine, with more lengths in between. We need to know about each of these links and the nature of the interactions between them to truly understand, and to intervene in, the rules of regulation on every scale.*"[25]

Carroll's work can be seen as an extension of James Lovelock's Gaia Hypothesis, the idea that Earth is a coherent system of life, self-regulating, self-changing, functioning as an immense organism with the capacity to keep itself healthy by controlling its physical and chemical environment. Lovelock never suggested that Earth was regulated "by foresight in the way of an intelligent animal" but still argued that it was alive and should be "thought of as a superorganism."[26]

Lovelock's and Carroll's ideas are byproducts of Charles Darwin's vision—which in turn was derivative of Alexander Von

Humboldt's and others—that speciation is not independent of the evolution of the material environment. Rather, the evolution of species and their environments are tightly coupled yet still evolve by natural selection. Along with the theory of evolution, is it possible that Gaia—by Lovelock's admission, "at the outer bounds of scientific credibility . . . as out of tune with the broader humanist world as it is with established science"—will end up being one of the more influential ideas in human history?[27] It's reasonable to hope appreciation for how Earth functions will lead to the recognition of its vulnerability and why living here requires us to function as part of something akin to a superorganism rather than as owner (capitalist), tenant (serf), or passenger (ignorant).

While it's unwise to evaluate the ramifications of species loss in only human-centered terms—every species has an inherent right to live—there are nonetheless any number of unknown human health benefits lost due to species extinction. For example, elephants' genetic instruction book includes about twenty copies of TP53—humans have but one—a gene that codes for a tumor-blocking protein that may explain why elephants have unusually low cancer rates. Poaching, hunting, and loss of habitat have led to a population decline from more than 25 million African elephants in 1550 to around 10 million by 1900, only 1.3 million as of 1979, and 415,000 today.[28] A keystone species in the African landscape, extinction will not only throw their habitat out of balance but could cause the loss of a treatment that would prevent untold human suffering and loss. At least 50,000 wild species are known to be used for medicine, energy, food, building materials, recreation, and Indigenous cultural practices, upon which 70 percent of the world's poor directly depend, emphasizing the ongoing losses are hugely consequential. And don't forget insects. Edward O. Wilson maintains insects "run the world" because they pollinate 87.5 percent of all plants. Every insect eats plants or eats something that eats plants. Without insects, humans would disappear in a few months. Invertebrate abundance (the number of insects) has declined 45 percent globally since 1974, a tragic statistic once you realize insects sustain Earth's ecosystems by sustaining plants. Have you noticed fewer insects on your car windshield?

THE SOIL LOSS AND DEPLETION CRISIS

In 2007 University of Washington Professor of Earth and Space Sciences David R. Montgomery wrote in *Dirt: The Erosion of Civilizations* that soil degradation and accelerated erosion were twin disasters that would determine our future. He maintains an "estimated twenty-four billion tons of soil are lost annually around the world, several tons for each person on the planet," making clear why soil loss, along with species extinction and atmospheric overheating make up the Trifecta Crisis.[29]

According to the UN Agricultural Agency, "Soil erosion affects soil health and productivity by removing the highly fertile topsoil and exposing the remaining soil [and] can also cause significant losses in biodiversity, damage to urban and rural infrastructure and, in severe cases, lead to displacement of human populations."[30] About a third of the world's soil has already been degraded. It takes a thousand years for Earth to generate three centimeters of soil, five hundred years to make an inch, and if current rates of losing and degrading the world's topsoil continue unabated, it could be gone within sixty years, meaning the end of farming. With 95 percent of our food a function of soil, functionally a nonrenewable resource, it's the basis of life. Like carbon, without soil, no life.

Despite this short sixty-year fuse, too much of the farm world plows ahead with the same antiquated methods that have been used since forever. Discing soil leads to erosion, the loss of soil's microbial life, and the release of soil's carbon into the atmosphere. As Rupa Marya and Raj Patel make clear, "Widespread deforestation and industrial agriculture are causing the loss of rich microbial diversity in soil, a living key to carbon sequestration and water retention."[31] Although farmers—primarily corporate monoculture farms—provide us with relatively inexpensive food, they are exacerbating the overload of atmospheric carbon and destroying natural soil exchange processes.

We are losing soil faster than it forms, but a given amount of soil is required to support each one of us, and our numbers are increasing. Earth needs erosion to keep refreshing the soil, but not so fast as to sweep it away altogether.[32] A combination of factors causes soil loss, degradation, and deterioration, including loss of fertile topsoil due to wind or water erosion, deforestation, extensive cultivation, deep plowing, poor agricultural practices like monocropping and improper crop rotation, overgrazing, fertilizer misuse, increased salinity, application of pesticides, and mining.

There is a circular connection between sunlight, vegetation, soil, and atmospheric carbon capture. The system is complex, with more than four parts, but these four elements are the basics and each is critical. Without sunlight plants couldn't photosynthesize. Without photosynthesis there would be neither oxygen for us to breathe, nor carbon for plant growth and subsurface life, nor the drawdown of atmospheric carbon. Again, no life.

The restoration on the Pitchfork Ranch is slowly rebuilding the soil, so much of it lost over the past several centuries. It will be a while before we can quantify the carbon content of the ranch's soil because the measuring process is sophisticated, undergoing development and commercial viability, but these calculations may be a part of our future participation in the carbon market. "May" is the operative word—that is, if the market can establish a definitive global threshold standard for high integrity credits that will provide buyers confidence and overcome the early 2023 *Guardian* report that said 90 percent of rainforest carbon credits purchased by Gucci, Disney, and Lavazza were 90 percent worthless.[33] This is an evolving marketplace, a venue for the exchange of carbon offsets as a means of meeting required emissions reduction with the eventual goal of being carbon neutral.

These waters are muddy. In *The Value of a Whale: On the Illusion of Green Capitalism*, Adrienne Buller maintains our current climate and nature crises solutions are grounded on the continuation of the same destructive processes, systems, and economic theories that not only delivered us to the fix we're in but have also delayed the necessary changes needed to address it. Trying to understand this profoundly human-caused crisis through a market-centric, growth-oriented economic lens has forestalled any meaningful solution to these crises. Buller's convinced the "green capitalist" model of climate cure is what Einstein was talking about when he cautioned against using the same kind of thinking that created a problem to solve it.

The reduction of the cattle numbers on the ranch has resulted in less grass consumption, meaning more carbon sequestration, a greater number of seeds for new plants, leaving more plant litter. Litterfall—dead plant material such as bark, leaves, and twigs—is required to make soil. Shallowing the landscape allows more water to seep in, which in turn fosters vegetation growth, more litter, and healthier soil. More growth means less wind erosion. Natural building of soil is immeasurable to the human eye, but it becomes evident by way of the increase in vegetation and color. Lucinda and I hardly notice the change, but visitors who return to the ranch after an absence of several years invariably comment on the change in vegetation density and color.

THE TRIFECTA CRISIS: A PLANET ON FIRE

With the planet on fire, there's plenty of blame to go around. And while we reasonably focus on the lawless behavior of the fossil fuel polluters, their enablers, and co-conspirators in Congress, there is room in this dilemma for science to take some of the blame. Despite the variation that accompanies climate crisis calculations, science has produced more than fourteen thousand peer-reviewed papers confirming that climate change is real, it has arrived, and it is human caused. Yet their computations invariably reflect the inherently cautious nature of scientists. Naomi Oreskes, historian of science at Harvard University and slayer of corporate myths, explains why some scientists arrived late to the crisis: "Scientists hold themselves to an extremely high bar before they are willing to say that they know something is true . . . [a] high level of confidence, which is the so-called 95 percentile confidence limit. . . . Scientists have been very, very afraid of crying wolf, and the consequence of that is that we've all been fiddling while Rome burns, or maybe I should say, we've been fiddling while Greenland melts."[34]

The fall of Rome and the climate crisis bear similarities for a number of people for a number of reasons; among the thinking is the inability of reason and science to pierce the veil of political ideology and religion-based faith. This flaw is something that also concerned Bruce Parker of the Center for Maritime Systems, Stevens Institute of Technology: "Would it be taking the point too far to suggest a parallel with the Romans, who kept the masses distracted from real-world problems by enticing them into the Coliseum to watch spectacles as gladiators battled to their deaths?"[35] And Irish American author Tim O'Reilly notes,

> As a former classicist turned technologist, I've lived in the shadow of the fall of Rome, the failure of its intellectual culture, and the stasis that gripped the Western world for the better part of one thousand years. What I fear most is that we will lack the will and foresight to face the world's problems squarely and will instead retreat from them into superstition and ignorance. . . . The Dark Ages were not something imposed from without, a breakdown in civilization due to barbarian invasions, but a choice, a turning away from knowledge and discovery into a kind of religious fundamentalism. . . . Civilizations do fail. We have never seen one that hasn't.[36]

Another obvious comparison of today's United States to the fall of Rome appears in classicist Edward J. Watts's 2018 *Mortal Republic: How Rome Fell into Tyranny*, which describes growing inequality, privatization, the powerful influence of money in politics, and leaders who consistently broke traditional norms and oversaw a political system that was no longer capable of reaching a shared concept of the common good. Watts explains how the Roman state "traded the liberty of political autonomy for the security of autocracy . . . [that] no republic is eternal," that Rome fell once politicians refused to adhere to "laws and norms" and citizens ignored leaders' "corrosive behaviors."[37] Former chief of staff to Secretary of State Colin Powell and now adjunct professor at the College of William and Mary Colonel Lawrence Wilkerson believes the United States has "become, in essence, the new Rome."[38] History is unambiguous that empires do fail.[39]

Few will openly quarrel with the covenant that everyone deserves an equal opportunity for the promise of the American Dream, an assurance now fully breached for the poor and those struggling to remain in the middle class. Growing numbers of Americans are slipping into the bleak, bloated world of debasing poverty. In these dark times, Americans find themselves represented by people of moral flexibility, callousness, and charm. We've entered an era of meaninglessness where money enables intellectual and moral poverty to persist. People are increasingly mad, mean, and inundated with misinformation.

Although not reassuring, the current "false facts" deluge—which can be thought of as a reality crisis or societal information disorder—has been experienced by humankind before. Johannes Gutenberg's fifteenth-century printing press revolutionized public access to information, just as the internet has shaken up the current era. Initially, as with today's internet, there were no standards requiring truth in publication; as a consequence, the printing press instigated thousands of deaths. Printed on the Gutenberg press, the 1484 tract *Malleus Maleficarum* (*Hammer of Witches*) swept across Europe and was said to lie on the bench of every judge and the desk of every magistrate. Claiming witches lurked everywhere, it was second only to the Bible in popularity and led to the burning at the stake of several hundred thousand people. Unconstrained by any obligation to fairness, accuracy, or the common good, the printing press wreaked havoc. Publications that disseminated false information caused Jean-Jacques Rousseau to call for banishment of printing.[40]

There is also something grievously wrong with Mark Zuckerberg's Facebook and the other social media platforms leading to widespread beliefs in oddball conspiracy theories and sudden appearances of strange groups and events. Hyper-partisanship, incendiary language, and identity-indulging tendencies have exploded online. While Jonathan Rauch's *The Constitution of Knowledge: A Defense of Truth* identified the internet as the cause of this problem, akin to the printing press, Max Fisher's *The Chaos Machine: The Inside Story of How Social Media Rewired Our Minds and Our World* pinpoints not what the problem is but how it works. The transgression or sin in these social media products is in that "the platform's algorithms and design deliberately shape users' experiences and incentives and therefore the users themselves."[41] Fine-tuned to users' clicks and engagements (Facebook, YouTube) and minutes and hours of view time (China's TikTok), researchers knew "our algorithms exploit the human brain's attraction to divisiveness" and that they were designed to deliver "more and more divisive content in an effort to gain user attention & increase time on the platform," a business model that keeps users glued to their schemes and increases profits:

> The early conventional wisdom, that social media promotes sensationalism and outrage, while accurate, turned out to drastically understate things. An ever-growing pool of evidence gathered by dozens of academics, reports, whistleblowers, and concerned citizens, suggests that its impact is far more profound. This technology exerts such a powerful pull on our psychology and our identity, and is so pervasive in our lives, that it changes how we think, behave, and relate to one another. The effect, multiplied across billions of users—

> ushering in a wholly new era in the human experience—has been to change society itself.[42]

The algorithms crafted by Meta (for Facebook, Instagram, and WhatsApp) and Google (YouTube) monetize humankind's draw to moral outrage by promoting hyper-partisanship, followed by a flood of misinformation. It's not clear why some of us are susceptible to the prodding of algorithms and others of us aren't, but enough of us are that our failure to come to grips with the severity of our environmental crises confirms that we're mired in a world of chaos. The United States has never faced an institutional crisis like the current one. Despite the short window to shift the world from the current trajectory of catastrophic decline, we await the creation of a system to regulate the electronic version of the printing press, distinguish between false information and true, verifiable knowledge, and forbid destructive, profit-driven algorithms.

Writing in the journal *Perspectives on Politics*, two of America's most respected political scientists, Martin Gilens of Princeton University and Benjamin Page of Northwestern University, paint a bleak picture of democracy in their paper "Testing Theories of American Politics." They found that the preferences of the average American have only a minuscule, "non-significant, near-zero" impact on public policy. Lawmakers respond almost exclusively to those with the most lobbying power and deepest pockets to bankroll their campaigns. The Gilens-Page "analysis indicates that economic elites and organized groups representing business interests have substantial independent impacts on U.S. government policy, while average citizens and mass-based interest groups have little or no independent influence."[43]

This conclusion was based on data from 1981 to 2002, before the Supreme Court opened the floodgates to big money in the *Citizens United* case, before the Wall Street bailout, before super PACs, and before the deluge of "dark money." Our democracy is now in even worse trouble; as of 2018, 85 percent of Americans thought government was run by a few big interests looking out for themselves, up from 29 percent in 1964. It seems we've stumbled into a pool of hopelessness when you add the 2021 Pew Research Center study finding that "two-in-three Americans agree with the phrase 'most politicians are corrupt.'"[44]

The median-voter theory, which holds that policy outcomes reflect the preferences of voters who represent the ideological center, has little application to today's America. In the face of Amitav Ghosh's view that the climate crisis is "perhaps the most important question ever to confront culture in its broadest sense," the Princeton and Pew studies' conclusions are horrifying.[45] The collapse of atmospheric function and the loss of biodiversity are driving those people most affected to extreme levels of grief and despair. The term *eco-anxiety*—chronic fear of environmental doom—has entered the lexicon. Experts warn that climate anxiety is on the rise, affecting huge numbers of young people around the world. An Oregon high-school counselor recently told me most students say they "will outlive the planet." Somehow these crises must be brought to the forefront of civic responsibility and government.

When climate breakdown, biodiversity collapse, and soil loss are considered in the context of fourteen thousand peer-reviewed science studies confirming these crises, we should be laser-focused on changes that will head off ecosystem collapse. Yet faced with the issue of our lifetime—the question of survival—the 99 percent appear to have no say. University of Edinburgh moral philosopher Elizabeth Cripps writes: "It seems unbelievable that we're not all chaining ourselves to the headquarters of oil and gas companies, or at least hammering on MPs' office doors."[46]

CHAPTER SIX

The Miracle of Habitat Restoration

"On this sand farm in Wisconsin, first worn out and then abandoned by our bigger-is-better society, we try to rebuild, with shovel and axe, what we are losing elsewhere."[1] Now is the time to do what Aldo Leopold—pictured with Estella in this chapter's wood engraving print—argued for in 1934: "The time has come for science to busy itself with the earth itself. The first step is to reconstruct a sample of what we had to start with." According to Curt Meine and Richard Knight, "Leopold drew upon a broad knowledge of botany, forestry, wildlife management, and ecology to restore biological diversity and productivity to degraded lands at [their] farm and elsewhere."[2]

Leopold had the foresight to practice and urge others to pursue restoring land more than eighty years ago. Habitat restoration as a discipline was unknown, yet he practiced it; his counsel is more relevant now than ever. Creating a slice of what was here to start with is what Leopold strove for. It's what we're working toward with the restoration on the Pitchfork Ranch and, with this book, encouraging others to pursue. This is regenerative change, replenishing the natural world: regreening, rewetting, rewilding, reforesting.

Our work over the past two decades has been a process of restoring the hydrological and ecological function of the land, providing habitat for wildlife and at-risk species, and sequestering carbon. As this land heals, it will capture increasing amounts of soil suspended in water flows and more carbon from the atmosphere. In 2019 the United Nations Intergovernmental Platform on Biodiversity and Ecosystem Services called for bold, far-reaching economic and social changes, including paying for large-scale ecological restoration of degraded lands. In keeping with the UN's call to action, this chapter demonstrates what each of us can do to capture and improve soil and draw down atmospheric carbon.

We've had access to resources aplenty—numerous books, articles, workshops, and conference proceedings—to explain *how* this work is done. If you live in the Southwest, Bill Zeedyk's *Let the Water Do the Work* is the best place to learn the mechanics of building grade-control structures.[3] Because the Pitchfork Ranch restoration is the work with which we're intimately connected, it's the restoration I can write about to demonstrate what others can do. Admittedly, some of our work has been accomplished with the use of a backhoe. But at least 80 percent of the more than one thousand restoration structures installed on the ranch are hand built. Hundreds of them were built by students from the Aldo Leopold Charter School in Silver City and by local young people hired with federal and state conservation grants, as well as our own funds.

When Aldo and Estella Leopold purchased their eighty-acre Wisconsin farm in 1935, they began doing what he called "reconstruction, restocking, rebuilding, doctoring sick land . . . [doing] some of the earliest restoration undertaken anywhere, help[ing] to lay the foundations for ecological restoration." He thought about damaged habitat as a "land organism" that was "sick" due to careless misuse. More than anyone, Leopold broadened the boundaries of conservation to include not just protecting habitat but restoring it: "Since Leopold's day the context and status of ecological restoration have evolved dramatically."[4] The term *restoration science* has entered the lexicon. His eldest daughter, Nina Leopold Bradley, wrote that her father believed "restoration could be a ritual of self-renewal. And so it was."[5]

Leopold had a full-time job as the first director at the University of Wisconsin's Arboretum, so he and his family, colleagues, and friends did their work at the "shack" on weekends. They restored the only structure on the land—a chicken coop made habitable for their family of seven—over the course of many years. Bradley recalls that her father "initiated a different relationship with the land, at once more personal and universal. From his own direct participation in restoration of the land, he was to come to a deeper appreciation of the ecological, ethical, and aesthetic understanding of land, and at the same time he was finding a dimension of his sense of place—so, too, did family members, colleagues in this venture."[6]

RESTORATION ON THE PITCHFORK RANCH

And so it has been for us. We realized early on that the learning curve for city dwellers transitioning to rural habitat restoration was steep and that it was necessary to hire or consult with the right people before tackling this work on our own. We learned restoring biological diversity and productivity to degraded lands requires an appreciation of scientific knowledge, adherence to a restoration protocol, persistence, and a long-term commitment to the land. We began the ciénaga restoration in 2005 by hiring someone with the skill to design the earliest stage of the project, a fortunate first step, as the initial layer of work has trickle-down implications felt yet today. We also felt more in tune, realizing that the ribbon of green trees was not merely a collection of individual Goodding's willows but an entwined collective, a resilient, complicated, and self-organizing ecosystem with a life of its own. There is plenty of scholarship to explain the communal aspect of forest and why people feel so much better in wilderness.[7]

Our basic restoration objective is to shallow the habitat and lengthen water routes. Grade-control structures modify degraded, incised stream channels by interrupting overly steep channel slopes, or incisions, slowing the movement of water. They capture suspended sediment in water flows and floods, over time filling the cuts and gullying caused by decades of unwise use. The idea is to help the damaged land reach a point where it can begin restoring itself. We intervene using active restoration in places that have been damaged beyond the point of self-recovery. The same idea applies to the ranch's entrenched 8.3-mile section of the Burro Ciénaga and the mile-plus incised historic ciénaga portion.

The Burro Ciénaga's main channel, side drainages, and arroyos are all treated with the same goal and similar design, shallowing the entire landscape (fig. 24). The base of incisions aggrades (fills in), becoming less and less deep. After the installation of grade-control structures, sediment in flows is captured

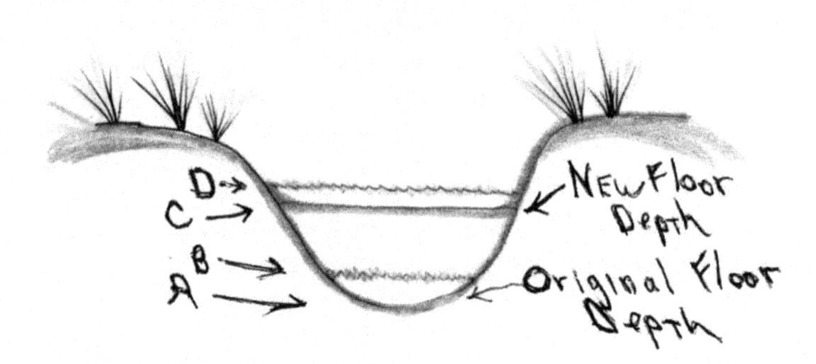

FIGURE 24 The process of shallowing the landscape, lessening the depth of incisions that form gullies or arroyos. A and B indicate the floor of the incision and the water level, respectively, before restoration. The floor of the channel, drainage, or incision is elevated ("shallowed") to the level at C, and the water level rises to D. Diagram by the author, 2022.

and drops to the floor. The floor of the channel, drainage, or incision is elevated ("shallowed"), then the water level rises. As self-healing shallows the incision further, water can access the floodplains, thereby capturing more sediment, improving soil depth and organic content, enhancing more vegetation establishment, and drawing down increasing amounts of atmospheric carbon.

HABITAT RESTORATION USING GRADE-CONTROL STRUCTURES

The Pitchfork Ranch lies in the heart of Apacheria, where the Apaches roamed for several hundred years before the arrival of Europeans. The translation of the Navajo term for these lands is "horizontal yellow" because the region was dominated by large swaths of grasslands; trees were kept at bay by fire, on average, every eight to ten years. The word *yellow* surely referred to the dried grass, and *horizontal* captured the dominant, historically shallow shape of the land. As we shallow the landscape, we are leveling, returning horizontal to the land, although it will never return to its native condition before the arrival of Spanish colonialists. No one alive knows what this land looked like then, as it has been altered and desiccated by human use, shifting much of the habitat to a new state. Nudging the land in the direction of its previous state is the goal, returning it to conditions that will slow water flow sufficiently to build soil and increase its capacity to capture and hold water, sustaining the grasses and other native vegetation that sequester carbon—and support economic activity. Quoting Leopold, "It is now generally understood that when soil loses fertility, or washes away faster than it forms, and when water systems exhibit abnormal floods and shortages, the land is sick."[8] It needs doctoring.

To be clear, ecological restoration is conscious work humans do in a landscape or ecosystem in order for it not only to resume *looking* like it did before damage caused by misuse but also to *act* and *function* as it did in its previous state. The goal is to replenish the system's self-organizing capacity for healing after its restoration jump-start. The restoration practitioner is engaged with the ongoing good news of fecundity, renewal, and promise of thwarting climate breakdown. Bill Zeedyk, who designed our work on the Pitchfork, suggests the most effective and cost-efficient way to repair habitat is to "let the water do the work."[9] He cautions, "There seems to be a mentality that the quickest way to do something is with a bulldozer. That's not necessarily true. You can change the course of a river by hand. So, I've been trying to develop techniques that empower individuals to try rather than say, 'It's not worth it to try.'"[10]

The grade-control structures pictured in this chapter are not meant to suggest a static suite of erosion control devices or a one-size-fits-all template; each location calls for its own type of intervention and structures must be modified to conform to the shape and damage of the site. These assemblages can be installed on any size property, and most can be built by hand. Habitats can be restored and carbon sequestered with grade-control structures just about anywhere: a larger piece of land like this ranch, land of smaller size in an urban setting, or other land being restored by one of the large numbers of habitat restoration organizations or groups that exist in almost every community.

Headcuts are the bane of healthy landscapes: they accelerate the erosive processes and, without treatment, unendingly erode the land until it eventually becomes water depleted and transformed into xeric conditions. Headcutting is a process where damaged conditions allow water to create a vertical nick or wound in the bed of the watercourse. As water flows over the nick, it undercuts upstream into the channel bed, scouring the channel, and "un-zipping" the watercourse up-channel and through tributaries. After the headcut reaches the top of the incision, at or near the location where the headcutting began, this erosive process can begin anew, repeating itself until the watercourse is no more.

Before we began restoration, there were a number of headcuts in the watercourse (fig. 6). Now there are none. Yet without help through restoration using grade-control structures, the land will never return to anything like the condition that existed before the damage. Headcuts will eventually create large gullies, carrying water faster, repeating the downcutting, and incising ever more deeply. Large drainages caused by this repetitive process are seen throughout the North American continent and the world at large.

The subtitle of Bill Zeedyk's book *Let the Water Do the Work, Induced Meandering, An Evolving Method for Restoring Incised Channels*, describes one of his most basic restoration tenets. His approach uses grade-control structures to nudge water into a back-and-forth meander throughout the channel—something undamaged watercourses do naturally—from right to left and back. We agree with Zeedyk's approach, although we have slightly altered the basic restoration template with less emphasis on induced meandering, prioritizing shallowing the landscape. We focus on filling incisions or shallowing the Burro Ciénaga, drainages, and smaller cuts in the land (fig. 25).

Although we use Zeedyk protocols, we prioritize soil aggradation rather than inducing water to zigzag, snake, or meander. Our core goal is to capture sediment suspended in water flows and raise the ground's incised surface level, shallowing the landscape.

We've built wicker weirs (fig. 25a) by hand, using one-seed juniper trees harvested on the ranch, trimmed into posts, sharpened, and sledgehammered into the channel bed. The posts were then handwoven with either juniper branches or cuttings from Goodding's willow trees. When this structure became full of sediment, topped out, and incapable of capturing any more soil and debris, we installed a second set of weirs atop the first, which was in turn followed by a third set after the second tier

FIGURE 25 Grade-control structures: (a) a set of three step-down wicker weirs, (b) rock dam, (c) post-baffle, (d) machine-built rock structure, one mile from the main watercourse on the lower portion of the ranch. Photographs by the author.

filled. Vegetation has now so overwhelmed this area we can't take a photograph that shows anything except a close-up of leaves and branches. This aggradation process will continue until the incised channel is closer to its former health and will eventually allow the ciénaga to reestablish itself, enabling the water to access its historic floodplain or terrace that has long been water starved. Another structure, a hand-built rock dam (fig. 25b), captures sediment, raising the channel bed. Around eight hundred of these structures have been installed in all but one of the thirty-four side channels that drain into the reach of the Burro Ciénaga on the ranch. Figure 25b depicts a structure ready to receive a second tier, having captured two feet of sediment. A hand-built post-baffle (fig. 25c) captures sediment and forces the flow leftward, harvesting sediment from down-channel left, depositing it below the structure to create a point bar and a meander, lengthening the channel.

Six machine-built structures have been installed in one of the larger side channels that drain into Burro Ciénaga on the ranch. After the tier 1 structure was installed (fig. 25d), sediment runoff after monsoon rainfall was captured and filled the drainage immediately up-channel from the structure. A year later, under the same grant, we installed a second tier atop the initial tier and extended the down-channel right arm. The 2017 monsoon rains completely filled the structure again.

Among many researchers who have worked on the ranch is a retired electrical engineering professor who studies moth species here. He and I measured the soil capture in the figure 25d structure at 700 tons behind the initial tier and another 360 tons after rains topped out the second tier. The weight of this amount of soil equates to about 650 elephants. With the 5 similar but smaller machine-built structures in this drainage, the total sediment capture is on the order of 5,000 tons. When you multiply that number by 33 for other side channels and the main watercourse, the total soil capture easily exceeds 100,000 tons.

The restoration protocol in *Let the Water Do the Work* reflects observation-based learning. Zeedyk worked in wildlife management—not just in the office but in the field—and retired after a thirty-five-year career with the U.S. Forest Service. He spent decades observing damaged landscapes and natural healing processes and integrated these observations and several practices of First Peoples into a restoration protocol consisting of a variety of grade-control structures that can be built by hand with materials commonly found on-site. The guiding principle of his strategy is the installation of structures or, when necessary, multiple tiers of grade-control structures to aggrade soil and stabilize an incision, enabling natural processes to eventually take over the restoration process, self-healing the habitat. These grade-control structures, whether in the side drainages or the main Burro Ciénaga watercourse, not only capture eroding soil but also slow water flow, increasing moisture infiltration into the soil, improving plant growth, and raising the water table.

Paired, same-location photography is a common way to monitor restoration and demonstrate improvements from habitat repair using these cost-effective, low-tech structures. Yearly, we take same-location, same-date photographs on the ranch that throw into relief the restoration changes over a period of years as a result of installing water-slowing, soil-capturing grade-control structures.

Photo point 10 (figs. 26 and 27), with Soldier's Farewell Hill on the horizon, is at mile three of Burro Ciénaga, so there are several hundred additional structures in the channel both up and down from this location. We installed a boulder baffle and planted Goodding's and coyote willow trees down-channel right. Both the structure and meander were obscured by expanded plant growth by 2020 (fig. 27).

At Bobby's Crossing, the incision with its two-foot vertical walls (fig. 28) has been filled with sediment (fig. 29). This is a typical consequence of soil capture by a Zuni bowl, a

FIGURE 26 Photo point 10, with Soldier's Farewell Hill in the distance. The arrow indicates the boulder baffle and Goodding's and coyote willow trees planted down-channel, inducing the left-coursing meander left of the arrow and between the vegetation on the left. Photograph by the author, 2005.

FIGURE 27 Photo point 10, with Soldier's Farewell Hill in the distance, same location. Photograph by the author, 2020.

FIGURE 28 Two-foot incision at Bobby's Crossing, just up-channel of photo point 10. Photograph by Lucinda Cole, 2007.

FIGURE 29 Bobby's Crossing, same location. A Zuni bowl was installed just below the pictured area. Photograph by Lucinda Cole, 2021.

boulder-constructed vessel or cup-shaped grade-control structure with a rock pour-over that Bill Zeedyk learned from First Peoples.

The dozen coyote willow trees that were pole planted with the boulder baffle at photo point 10 (fig. 26) are now more than a thousand in number, trees migrating both up and down channel by way of rhizomes, modified subterranean plant stems that send out roots and shoots from their nodes (fig. 31).

The upper one and a half mile of the ranch's reach of the deeply incised Burro Ciénaga (fig. 32) now remains wet throughout the year. Despite the new vegetation and sediment capture (fig. 33), this partially restored reach of the ciénaga remains deeply incised, even though the flow has slowed considerably and the wound, both bottom and sides, is continually aggrading, thereby raising, narrowing, and shallowing the incision.

Woody flood debris has been captured by the Goodding's willow trees (fig. 34), planted as part of the 2005 restoration (fig. 32). This beaver-like dam remained in place for several years, continuing to capture more debris and sediment. An inordinately large flood in 2022 dislodged much of this captured debris, while the formerly hinge-felled, regrown Goodding's willows remained.

About a quarter mile down-channel from the eight-foot incision in the Burro Ciénaga (figs. 32–34), a restored riparian area (figs. 35 and 36) demonstrates the benefit of grade-control structures, soil moisture infiltration, and vegetation recruitment and makes plain the benefit of removing cattle from riparian areas. With less than 2 percent of riparian habitat remaining in the entire Southwest, fencing cattle off these long-abused, tender areas is a fundamental restoration protocol—no exceptions.

Although ownership (whether federal, state, or private) makes a difference for funders, it doesn't determine the location for placement of grade-control structures for us; rather, damage to the terrain and areas of soil loss dictate where structures are installed. Most of the grade-control structures

FIGURE 30 A view of midchannel, immediately down-channel of Bobby's Crossing. Photograph by the author, 2005.

FIGURE 31 A view of midchannel, same location. Photograph by the author, 2017.

FIGURE 32 An eight-foot incision (down-channel, right) in Burro Ciénaga, immediately after the initial restoration work. Photograph by Lucinda Cole, 2005.

FIGURE 33 The healing incision in Burro Ciénaga, same location. Photograph by Lucinda Cole, 2017.

FIGURE 34 Lucinda stands with debris captured by willows at the grade-control structure. Photograph by the author, 2018.

are either in or near the riparian area or the side channels that drain into the Burro Ciénaga. Less steep channel incisions after restoration and shallower side drainages reduce rainwater runoff and soil erosion. Moisture infiltration and wicking increase, sustaining the expansion of grasses and other vegetation. These processes reinforce one another, further slowing water runoff, promote additional vegetation growth, and increase carbon capture. Although we are years away from full ciénaga recovery, once restoration enables sufficient sediment capture and vegetation establishment, self-healing can begin; our restoration work will lessen. Self-healing is already occurring through most upper reaches, and in others the growth of in-channel vegetation is so thick that it prohibits instillation of additional structures. Having observed two decades of

FIGURE 35 Restored riparian area where cattle grazed. Photograph by Lucinda Cole, 2005.

FIGURE 36 Restored riparian area after removal of cattle, same location. Photograph by Lucinda Cole, 2010.

change and growth as of this writing, we now realize the incised ciénaga may not fully recover during our lifetime. Someone else may need to step up and complete the task.

HABITAT RESTORATION SOLUTIONS ARE AVAILABLE TO EVERYONE

Beyond transitioning to sustainable fuels and reducing consumption, habitat restoration—which can be thought of as "climate restoration"—may be the most important personal climate mitigation strategy you can adopt. It's the activity that has dominated my and Lucinda's lives since 2005 and will hopefully become a part of yours. Considered an adaptation and mitigation strategy, habitat restoration is something everyone can do. It can be carried out almost anywhere, on property of any size, irrespective of ownership. *Everyone* can pursue restoration, and it can be done on *any* size property, just about *anywhere.*

Because of the climate crisis and the need to draw down excess carbon, restoration has become a more important part in climate crisis solutions, significantly beyond the goals Lucinda and I established when we began this work. With the 2023 atmospheric carbon count hovering around 424 parts per million (ppm) and given the need to reduce that count to at least 350 ppm, habitat restoration, more vegetation, and more photosynthesis by which plants capture and inject carbon back into the soil are the best currently known ways to draw down atmospheric carbon. Beholden to technology, engineers in corporate America are working to invent machines to do what nature does for free (and to make money doing it). There are even suggestions we can harvest lunar dust for an atmospheric umbrella to lessen solar heat. As Einstein warned: "We cannot solve our problems with the same thinking we used when we created them."[11]

With so much concern focused on ending the use of fossil fuels, transitioning to a fossil-fuel-free economy, and reducing emission of greenhouse gases, it's easy to forget about the excess carbon already in the atmosphere. The saving grace, the solution to drawing down more than one-third of the atmospheric carbon overload, is straightforward, easy to administer, and can be inexpensive: stop clear-cutting and felling rainforest for grain feed; plant trees and other vegetation; restore and expand riparian and other vegetated areas; start a climate garden; or build a wet spot.

Increasingly, many communities have neighborhood gardens where area residents can get their hands in the dirt and grow healthy food. The shrink-your-yard, restoration, and climate garden strategies proposed here form an action plan with five goals: reducing water misuse; gathering water and building soil to grow plants; sequestering atmospheric carbon; growing your own food, thereby reducing carbon dioxide emissions caused by long-distance food transportation; and planting tree seedlings. This action plan is one in which we all can participate to help save the planet, our lives, and the lives of Earth's other inhabitants.

People with dissimilar lifestyles, living in differing surroundings and on different-sized properties, are also restoring habitat near us (table 1). Many thousands of people and hundreds of organizations are engaged in similar landscape repair throughout the United States. The same restoration techniques used on this ranch can be used equally on vastly different-sized properties, both urban and rural, and a wide range of habitats. Virtually any residence—irrespective of owners' income levels and socioeconomic differences—can be a setting for habitat restoration.

Each of the restoration projects near us (table 1) have transformed degraded land into a depositional habitat: soil moisture increases; erosion is reduced; native plant life returns or increases; all manner of animals return to these refurbished landscapes; temperature is reduced; and a contribution to ameliorating the climate crisis is made by sequestering carbon.

TABLE 1 Properties of various sizes and types that have benefited from restoration practices.

OCCUPANTS	SIZE	DESCRIPTION	AREA IN ACRES	PROPERTY
John Doe	Micro	Urban Rental	0.02	City Lot
Ann Hedlund	City Large	Urban Home	0.36	8 City Lots
Donna Stevens	Rural Medium	Rural Home	11	Residence
George Farmer & Linda Zatopek	Rural Large	Axel Canyon Preserve	83	Ecological Preserve
A.T. & Lucinda Cole	Larger	Pitchfork Ranch	11,393	Ranch
Josiah Austin	Huge	La Cienega Ranch	70,000	Ranch

- **John Doe's Urban Rental**. No property is too small to be restored, as is apparent from John Doe's 0.02-acre rental residence in Silver City, New Mexico. This property has benefited from curb cuts where rainwater from the street is diverted onto the property. John has also established berms for water retention and added plantings that have improved water retention and increased carbon sequestration.
- **Ann Lane Hedlund's Urban Residence**. Ann Lane Hedlund moved to Silver City in 2008 and purchased the 0.36-acre Silver City Spring Street residence in 2012. Her restoration included removing the water-hungry grass lawn, midwestern flowers, evergreen hedges, and other nonnative vegetation and replacing them with local, drought-tolerant plants—native mulberry, Emory oak, crimson sage, soap tree, columbine, and giant sacaton grass harvested from the Pitchfork, plus seventy other plants, all indigenous to the Gila River region.

Both gray water from the house and a water-harvesting feature from the street—with the help of city authorized curb cuts and volunteers from the local Youth Conservation Corps—are in place to keep two immense Arizona cypress trees, planted circa 1906, and other native vegetation healthy. Sharing six four-by-eight-feet garden beds with eight neighbors, she and others nearby are growing some of their own food. As is so often the case in healthy communities, good ideas are infectious, encouraging nearby residents to landscape their homes differently and in ways that will conserve water.

- **Donna Stevens's Rural Residence**. Donna Stevens purchased her eleven-acre rural homestead twenty-five miles south of Silver City in 1985 and has made strides to arrest erosion and increase rainwater capture. As with most habitat restoration enthusiasts, Donna initially focused on repairing the habitat for its own sake and for the wildlife that are dependent on the land. And while those goals remained important, as the burden of climate overheating became better understood, she too began to think in terms of carbon sequestration, a transition made easy because the tasks of habitat repair and carbon sequestration are identical.
- **Axel Canyon Preserve**. George Farmer and Linda Zatopek retired to southwest New Mexico from Texas in 2006 and began restoration on their dream Axel Canyon Preserve. Similar to the restoration goals for these other five properties, their objective is to improve the wildlife habitat by cultivating vegetation and enhancing the diversity of plants growing in and near the drainages and alluvial fans on their eighty-three-acre property, now a wildlife preserve. Their efforts have slowed the impact of sheet flow and gully erosion from surrounding uplands, induced sediment deposition, stabilized eroding banks, and increased carbon sequestration.

  The practices involved in their restoration work include installation of wicker contour weirs, brush channel lin-

ers, and one-rock dams in tributary arroyos, as well as construction of plug-and-pond systems to restore channel headcut gullies. Native plants have been transplanted within the brush baffles. Contour weirs were also used to stabilize eroding banks and induce sediment deposition, including seep willow, coyote willow, Goodding's willow, cottonwood, false indigo, four-wing saltbush, and giant sacaton grass harvested from the Pitchfork Ranch. The various plantings have dispersed water flow, captured sediment, improved the soil, and created improved habitat where it had been severely damaged in decades of misuse.
- **Pitchfork Ranch**. The restoration work at our 11,393-acre Pitchfork Ranch is similar: pole planting of Goodding's, coyote willows, and Fremont cottonwood trees; installing more than 200 grade-control structures in the reach of the Burro Ciénaga on the ranch; building 800 restoration structures in 33 of 34 side channels. We've planted and propagated hundreds of trees and other native plants, which in turn have recruited hundreds more, and fenced cattle off riparian areas so that ash and Goodding's willow trees can replace cattle-grazed scrub-brush-like trees (figs. 26 through 36).
- **La Cienega Ranch**. The changes to Josiah Austin's La Cienega Ranch are similar to the Pitchfork Ranch in that Austin's focus is not only on preservation but also habitat restoration, using grade-control structures: gully plugs, what he calls leaky weirs (rock and cement structures that allow water to sift through) and loose rock structures that also slow water and aggregate suspended sediment in water flows.

Three aspects are common to the habitat repair in these examples: landscape function, adaptation to the Trifecta Crisis, and drawing down and sequestering atmospheric carbon. On all six properties, water runoff is slowed, wicking and seepage increases, the water table benefits, and more of the water that falls on the land remains. Soil erosion is minimized and even reversed. These restoration projects also serve the common good by providing ecosystem services: clean water, reduced erosion, and fire protection. The three larger projects are examples of the large-scale conservation efforts proponents of natural climate solutions are calling for. There are efforts to establish a new national market to pay agriculturalists for drawing down carbon to their soil, improving water quality, and other ecosystem services; BCarbon in Texas is among the best. The work on these six properties and the potential for ecosystem services will hopefully influence neighbors and others.

There are other important opportunities for habitat restoration beyond what we are doing on the Pitchfork Ranch and what our nearby neighbors are doing. Responsible, climate-mitigating residential land use can become a part of the transition out of the current degraded state of much of the Southwest. Not many Americans alive today recall victory gardens, but in 1941, 75 percent of the 137 million people in the United States had access to these gardens, and 42 percent of fresh vegetables came from backyards. Born of necessity due to World War II, suddenly, there were more than 19 million gardens in America. If the same 75 percent of today's 310 million U.S. citizens developed climate gardens, the more than 19 million victory gardens of 1941 would more than double to 50 million climate gardens. The most important learning from this nationwide World War II program is that it demonstrates social mobilization of individuals and families on a massive scale can be successful. It can be done again.

If you don't have the energy or space for a climate garden, join a community garden or create a rich, wet backyard sweet spot, a pond, marsh-like area with native plants, a way to personally draw down atmospheric carbon.

Ohio State University Distinguished Professor of Soil Science Rattan Lal emailed, "If there were two billion such gardens in the world, it would make an incremental reduction in carbon while also advancing global food and nutritional security. The latter part is critical because the world has 800 million people who do not get enough calories and another two billion who are suffering from malnutrition (deficiency of protein and micronutrients)."[12]

The 2019 Children's Defense Fund report found that 12.8 million children in America—about one in five—are living in homes with income below the poverty line, depriving them of proper nutrition. Climate gardens would not only provide inexpensive and much-needed safe, high-quality food but would create biodegradable straws (plants) to draw heat-trapping carbon dioxide out of the atmosphere, disperse it in the soil, and help arrest global overheating.

Decades ago, as a student at Indiana University's School of Public and Environmental Affairs, Silver City's Mike Fugali scheduled a visit with Lynton K. Caldwell, professor emeritus, who founded the college. Caldwell was the principal architect of the 1969 National Environmental Policy Act, the first of its kind in the world. Fugali wanted to know Caldwell's thinking about where the world stood in terms of damage to the environment and what he could do about it. Caldwell told him Earth was at a critical planetary threshold and to "study restoration ecology because it's the only kind we have now."[13] In keeping with Caldwell's counsel, the late Barry Lopez got it right in urging the Voice of the Streets when he wrote, "We can no longer afford to carry on in a prolonged era of polite reflection and ineffective resistance."[14]

CHAPTER SEVEN

Healthy Soil

Hope Beneath Our Feet

WHILE WRITING ABOUT THE causes and extent of this Trifecta Crisis, I found myself sinking into a dispirited outlook of "why bother?" Why should we trouble ourselves to restore the ciénaga and grasslands, introduce at-risk species, pursue carbon sequestration, and write about their importance if we're doomed to face temperatures above levels at which humankind can survive?

In a conversation with ornithologist Carl Bock, who with his spouse, Jane Bock, often visits the ranch and helped establish the science portion of our work, I found relief from my worry; even if humankind perishes, other life will persevere, and the ciénaga restoration efforts will serve them. This land will provide a refuge for those survivors even if we reach a time of human extinction. Most people think of this Trifecta Crisis as a human problem when it's a planetary, all-life-forms dilemma. Earth will get along fine without us.

Just in time to encourage a grassroots movement, recent science and news reporting maintain natural climate solutions are among the best ways to individually address these crises. Natural climate solutions are "conservation, restoration and improved land management actions that increase carbon storage and/or avoid greenhouse emissions." Habitat restoration is one of twenty-one categories of natural climate solutions, centering us in the places where we live. Studies outline "the potential for [natural climate solutions] to address the converging crises of climate change and nature loss, while also helping to deliver sustainable development in line with the United Nations Sustainable Development Goals—providing equitable livelihoods, advancing education and equity, and improving natural resource management."[1]

My overcast of doom cleared further when I realized the number of scientists and scholars who are optimistic that there are ways to counter global overheating by employing a variety of natural climate solutions, new economic models, and existing technology. We don't need to invent anything other than a new outlook, a new, more measured lifeway. It's worth emphasizing:

hope is neither an emotion nor idealism, and it's not simple optimism. Hope is a discipline, a behavior, an intuition, a practice, a choice. It is a long-term plan grounded in a commitment to the future, a time to come where change is earned by engagement, political agitation, and disruptive and forceful action in taking to the streets and giving voice to the just.

Hope is in the ground beneath our feet. Each of us can make a measurable difference by working with soil, turning our yards and other lands into carbon-sequestration gardens, and becoming drivers of biodiversity and human health. People we might think of as "soil enthusiasts" frame their climate crisis solutions in a variety of ways; in its simplest form, and cribbing the title of Kristin Ohlson's book, "soil will save us."[2] Think of soil as survival. Growing numbers of scientists, scholars, writers, educators, farmers, ranchers, land managers, gardeners, chefs, restaurateurs, and other enthusiasts are hopeful, rejecting the doomsayers who believe we have already passed various tipping points beyond which survival is no longer possible. These hopefuls see a way to avoid the accelerating trajectory toward ruin by identifying the science, politics, and strategies for abandoning the fossil fuel merry-go-round, focusing on soil as part of the template for a net-zero future.

Yet it's hard to know our future. On the one hand, in 2022, global fossil fuel emissions were on track to reach record highs with no indication of declining, while on the other, the cost of solar power and lithium-battery technology has fallen more than 85 percent and the cost of wind power is down by more than 55 percent. Solar power is on track to be the cheapest source of electricity in history, with 90 percent of the global population living in places where new renewable power will be less costly than new dirty power.

Despite today's many politicians compromised by fossil fuel subsidies, there is a deep history of land loyalists in America's past. Soon after nationhood, Patrick Henry proclaimed, "Since the achievement of our independence, he is the greatest Patriot, who stops the most gullies."[3] When Franklin Roosevelt signed the first soil and conservation legislation into law, looking deep into the past, he stated, "The history of every nation is eventually written in the way in which it cares for its soil."[4]

One of the reasons the United States has accomplished so little in addressing accelerating greenhouse gas emissions and companion crises is what some historians refer to as a "stunning historical blankness" caused by our lack of interest in history and the internet-induced "one-touch knowledge" of "presentism," a condition where "the total sum of knowledge of anything beyond the present seems to be dwindling."[5] There are those who believe "we are in danger of becoming mere knowledge tourists, hopping from attraction to attraction at 30,000 feet without respecting the ground that lies between. . . . [A] dimension of intelligence is disappearing."[6] Mass forgetting is the collective amnesia in which presentism dooms us to repeat history without noticing it. It bears repeating: the unexamined past always forces its way into the present.

As soon as we began to leave the hunter-gatherer way of life, humans initiated the destruction of Earth's soil.[7] The innovation of agriculture caused the human population to skyrocket from five million to eight billion people in a mere ten thousand years, just a small portion of the time *Homo sapiens* have been on Earth.[8] This exponential growth—along with the disc-harrowing methods of consolidated corporate agriculture and other destructive ways in which we produce food, use fossil fuels, and increase our resource consumption—underlie the deterioration of the world's ecosystems.

Journalist Kristin Ohlson's book *The Soil Will Save Us* is among a growing number of writings chronicling the thinkers and practitioners who contend we can help reverse climate change by restoring soil. In some areas of the world where cultivation has been going on for millennia, soil carbon depletion is up to 80 percent or more. The world's soils have lost up to 80 billion tons of carbon.[9] For eons, this carbon was underfoot, but now it's floating in the rapidly overheating atmosphere. Scientists tell us more carbon resides in soil than in the atmosphere and all plant life combined; there are 2.5 trillion tons of carbon in soil compared to 800 billion tons in the atmosphere and 560 billion tons in plant and animal life. For drawing down excess atmospheric carbon, natural climate solutions are the only viable crisis remedies: simple, inexpensive, and well defined, with an exceptional record of documented successes. These 21 natural climate solutions illustrate the famous principle of simplicity, Occam's Razor: "All things being equal, the simplest solution tends to be the best one."[10]

Proposed geoengineering solutions to this Trifecta Crisis require scrupulous evaluation. One of many examples is direct air capture, an industrial technology that functions like giant vacuum fans to magnetically draw carbon from the atmosphere and inject it into the ground for permanent storage. Although it's an option not far beyond the experimental stage, it typifies the thinking that got us into this mess. Similarly, rather than pursue technological solutions to the production of power, why not focus on design? Stanford's Amory Lovins, said to be the "Einstein of energy efficiency," maintains that "integrative, or whole-system design" logic in factories, vehicles, equipment, and buildings would result in a severalfold energy savings. Potential solutions to these crises are endless, yet we have so little time.

Increasing numbers of practicing soil enthusiasts—calling themselves soil farmers, microbe farmers, carbon farmers, and soil or carbon ranchers—and various supporters of these soil loyalists are at the trailhead of the path to restoring the land, air, and water, as well as the species that live there, helping arrest the climate catastrophe. They are focused on the interaction between soil and climate, between soil carbon and global warming. Rattan Lal, director of the Carbon Management and Sequestration Center at Ohio State University, maintains if we arrest carbon emissions, responsible soil management could recapture most of the misplaced atmospheric carbon by bringing soil back to health and simply allowing plants to do what they have always done: photosynthesize.[11]

In elementary school we learned how photosynthesis converts sunlight and carbon dioxide into oxygen for us and other animals to breathe and carbon for plants to consume. I don't recall a thing about plants injecting carbon into soil. This is among this book's more important ideas: natural climate solutions accelerate carbon drawdown from the atmosphere through practices that return it to the soil by way of photosynthesis. It's a natural process capable of drawing down our legacy load of atmospheric carbon back to Earth, where it serves our future, rather than ends it.

PHOTOSYNTHESIS: OUR GREAT GREEN HOPE

When sunlight reaches vegetation, the plant (or photosynthesizing algae) absorbs atmospheric carbon dioxide. Carbon dioxide is a naturally occurring chemical compound in which two oxygen atoms are bonded to a single carbon atom. The plant absorbs the carbon dioxide via photosynthesis, then breaks the carbon apart from the oxygen atoms, consumes the carbon, and returns the oxygen to the atmosphere for us to breathe. The plant then converts the remaining carbon into high-energy sugars and carbohydrates. As Kristin Ohlson writes, "The prize is carbon. . . . The carbon sugars created through photosynthesis are the building blocks of life—they are the beginning of the food chain for just about everything on earth."[12]

Once converted, some of this newly minted carbon feeds the aboveground portion of the plant, while up to 40 percent of the carbon goes underground and migrates into the soil, providing sustenance to soil microbes. The microbes—bacteria and fungi—secrete enzymes that liberate minerals from Earth, which in turn provide plants with trace minerals and nutrients inaccessible on their own. Bacteria, which are ingested by protozoa and nematodes, can then consume life-giving minerals that otherwise could not be used by plants or micro-arthropods near the plant's roots. These organisms excrete the consumed

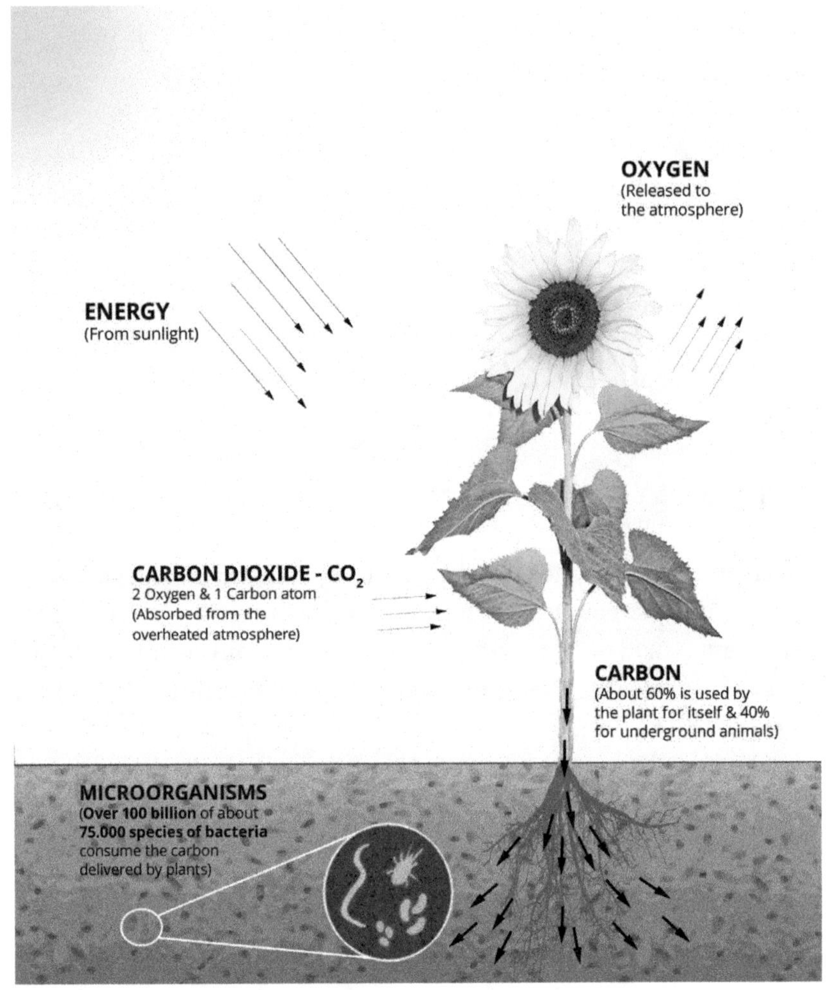

FIGURE 37 The photosynthesis process supplies oxygen for us and other animals to breathe and carbon for the plant itself and for organisms living in soil. Diagram by Cristián Uribe, 2021.

byproducts, and only then are these minerals in a form that plants can use. Without giving it a moment's thought, we walk and drive daily over this vast and sophisticated underground kingdom of untold trillions upon trillions of microorganisms without which our lives would not exist.

In terms of planetary health, it's helpful to think of each blade of grass as a straw, pulling carbon out of the air and into the plant's tissues, where it is converted to sugars and piped into the soil. The carbon in carbon dioxide is critical for the life of plants, just as the oxygen serves its life-giving property for humans. Without plants, we would simply run out of oxygen. Thinking of the blade of grass as a straw complements the similar figure of speech of a blade of grass as a dam in the context of habitat restoration. We know when water flows down-slope across the land surface or down-channel, suspended sediment bumps into vegetation, and some of it remains. This process lessens erosion and adds more sediment to the land surface, thereby aggrading the land to a deeper, richer soil for plant life and a shallower land surface with less erosion. Each blade of grass provides this two-for-one benefit: building soil and sequestering carbon.

The U.S. Geological Survey has surveyed and mapped the United States to identify which habitats can best store carbon and has explored ways to encourage this capability by bettering our stewardship of natural resources. The goal is to give land managers and policy makers tools to improve habitat and to increase the capture of carbon emissions from the atmosphere. It's one thing to *stop* carbon emissions and another to *draw down* the legacy load of carbon already in the atmosphere. The soil solution for combating climate change received a significant boost when a consortium of nineteen universities and other organizations released a paper published in the *Proceedings of the National Academy of Sciences of the United States* titled, simply, "Natural Climate Solutions." The 2015 Paris Agreement was approved by 197 nations, who committed to keeping the increase in the global average temperature at well below the variance of two degrees Celsius above preindustrial levels (preferably, below 1.5 degrees Celsius). The study contends "the most mature carbon dioxide removal method is improved land stewardship . . . better stewardship of land is needed to achieve the Paris Agreement goal of holding warming below 2°C."[13]

The National Academy of Sciences report identified and quantified twenty natural climate solutions, that is, "conservation, restoration and land management actions that increase carbon storage and/or avoid greenhouse gas emissions across global forests, wetlands, grasslands and agricultural lands." This may be the first time a specific number has been attributed to this potential: "natural climate solutions can provide up to 37% of cost-effective CO_2 mitigation needed through 2030 for a [greater than] 66% chance of holding warming below 2°C."[14]

Furthermore, these practices "offer water filtration, flood buffering, soil health, biodiversity habitat, and enhanced climate resilience," and "nature has the potential for large additional climate mitigation by combining enhanced land sinks with reduced emissions." This approach translates the Paris Agreement "climate commitments into specific natural climate solution actions that can be taken by government, private sector, and local stakeholders." Per unit area, wetlands like the ranch's ciénaga can hold the highest carbon stocks and greatest hydrologic ecosystem services, including climate resilience, "offering 14% of [natural climate solution] mitigation opportunities needed to hold warming below 2°C." Regreening, reforestation, and rewilding "the planet through conservation, restoration, and improved land management is a necessary step for our transition to a carbon neutral economy and a stable climate."[15] These practices are inherent in Wallace Stegner's "geography of hope" and actually address more than the Trifecta Crisis, including persistent, ever more egregious social and economic inequality.

In 2018, another natural climate solutions paper—this one for just the United States, rather than the entire planet—was published in *Science Advances.* It refines the earlier paper's coarse global analysis and expanded and updated the range of options, saying, "natural climate solutions—21 conservation, restoration, and improved management interventions on natural agricultural lands— . . . are the most mature approaches available for carbon conservation and uptake . . . [and can] increase carbon storage and avoid greenhouse gas emissions . . . [up to] 21% of the current net annual emissions of the United States."[16] Natural climate solutions entered the mainstream in early April 2019 when *The Guardian*, followed by the *New York Times*, *Washington Post*, and other news outlets, carried stories touting natural climate solutions. There needs to be a resurgence of interest in funding for soil health practices, including organic-amendment applications to rangelands and parallel soil improvement efforts in croplands.[17]

In addition to mapping the United States to identify which areas can draw down the most carbon, the U.S. Geological Survey has also initiated studies measuring the efficacy of grade-control structures, quantifying the outcomes of the features we and others are installing to restore habitat. Its Aridland Water Harvesting Study, the first to scientifically quantify *all* the benefits and down-channel outcome of grade-control structures, compares them to beaver dams and their analogs. This group of scientists is committed to quantifying with specificity data, validating the work of habitat restorationists, and hopefully causing the allocation of greater funding to this practice and other natural climate solutions.[18]

THE DESTRUCTION OF SOIL

Eliminating the bulk of greenhouse gas emissions and the atmospheric legacy load of our fossil fuel addiction is a Herculean task. Among other changes, this process requires replacing the abusive farming and cattle-raising practices that have depleted the landscape. More than eight thousand years of agriculture—forest clearing, irrigation, and plowing—is a story of agribusiness according to the model or customs of tilling, monocropping, keeping of feeder cattle, cattle overstocking, poor grazing practices, and poisoning the land and animals with pesticides, herbicides, hormones, antibiotics, and chemical fertilizers that not only kill microbial life in soil but cause all manner of human health ills.

Characterized by deep plowing and disc harrowing, modern mechanized farming damages the soil structure by disrupting the crucial underground networks of mycorrhizal fungi and splintering the soil aggregates that retain water and gases in the soil. Tilling breaks the earth, piercing delicate root systems and causing plants to release more food than normal, throwing their systems off-balance. This in turn exposes and releases soil carbon into the air, where it combines with oxygen and drifts into the atmosphere as carbon dioxide. There is no comparable process in nature that repeatedly and regularly turns over soil at the depths to which land the world over is being plowed, so neither soil organisms nor plants have evolved or adapted to this extreme and unnecessary disturbance.

The current farming practice of plowing-induced soil fracturing creates barren land that starves microbes in the soil. Much of the soil in the United States has simply lost its life-giving carbon due to plowing and poor land management. Poor soil results in fewer plants; fewer plants mean less photosynthesis; less photosynthesis means less carbon sequestration; and less carbon sequestration results in a hotter atmosphere. Farmers, ranchers, land managers, and the 80 percent of homeowners with yards must emphasize native vegetation and successful plant recruitment to fill in barren earth: photosynthesis doesn't occur on bare ground.

THE LIFE-GIVING UNDERWORLD OF HEALTHY SOIL

Plants distribute carbon sugars to microorganisms through their roots in exchange for nutrients, but when artificial fertilizers are added to the soil, they disrupt a natural trading system that has been established over eons. Soil organisms simply can't reach their natural food supply without access to the roots of healthy plants. Summarizing the work of U.S. Department of Agriculture microbiologist Kristine Nichols, Kristin Ohlson writes, "Without their carbon meal, the mycorrhizal fungi—[that] can shoot out their hyphae up to 250 yards and connect to different plants at once—can't grow and stretch their strands of carbon through the soil. They and other soil microorganisms can't produce the glues that fix carbon in the soil and build the aggregates that hold water."[19] Starved of natural processes, mycorrhizal fungi go dormant and often die, resulting in increasingly unproductive and depleted soil, prompting the farmer to add more fertilizer and worsen the cycle of destruction. (See appendix B for a more detailed discussion of mycorrhizal symbiosis.)

Loss of this most fundamental exchange in nature—a process that might be thought of as nature's opus—leaves the soil so depleted and non-adhesive as to be unproductive and unprofitable for food production. Aggregates develop in properly cared-for and vegetated soil, and the dirt becomes spongelike, holding water that supports plant life even when rain is sparse. These advantages of healthy soil can be folded into the newly designated group of benefits referred to as "ecosystem services"—ecosystems that service us. In the last several decades, the identification of essential nutrients in this remarkable subsurface exchange system has ballooned from a mere three known nourishment-providing substances to thirty-two nutrients. As the rapidly emerging field of soil science gains traction, the discovery of nutrients is expected to continue. These soil facts should unhinge one's sense of humankind's significance: there are more microorganisms in a cup of healthy soil than the number of all the humans that have ever lived (113 billion people). The dark underground world is thought to account for up to 95 percent of the planet's species diversity.

This soil food web—the rich, complex, and delicate ecosystem just below Earth's surface—is a food chain or food cycle that has the same flow as the familiar human system of eating, digesting, defecating, and decomposing. This everyday process is actually the cycle of all living things, whether in the above-ground animal world or in the subsurface microbial ecosystem. As scientists and others look for solutions to exhausted soil, pollution, food scarcity and water shortages, and climate breakdown, the world beneath our feet has become top priority for many. Carbon is the key. Internationally renowned Australian soil ecologist Christine Jones explains: "Mycorrhizal fungi . . . acquire their energy in a liquid form, as soluble carbon directly from actively growing plant roots. By this process they are actively drawing down atmospheric carbon and turning it into humus, often quite deep in the soil profile, where it is protected from oxidation."[20]

NEW PRACTICES FOR HEALTHY LAND STEWARDSHIP

Writing about climate crisis solutions in *Grass, Soil, Hope: A Journey Through Carbon Country*, Courtney White and Michael Pollan survey the strategies of a growing number of farmers, ranchers, and others who are pursuing new approaches to food production and land stewardship—dubbed "regenerative agriculture," "new agrarianism," "conservation agriculture," and "agroecology"—as a way not only to grow food but also to build soil, store carbon, and reverse the warming of the climate that is overwhelming the planet. These new practices are important because current anti-environmental agricultural practices contribute about a third of all greenhouse gases entering the atmosphere.

In contrast, these new regenerative farming and ranching practices can begin to return carbon to the soil. They emphasize soil as the solution that will help solve the climate crisis and offer a host of practices—sustainable methods based on a combination of ancient agricultural traditions, modern science, and local knowledge—that serve as commonsense answers to the need for food security and to arrest climate overheating.

- **Pasture cropping**: This no-till drilling method bucks agricultural tradition by using the same land for both crops and pasture and can create higher counts of fungi and bacteria, more than a 100 percent increase in minerals and nutrients, and more than a 200 percent increase in both carbon and water-holding capacity in just a decade of the practice.
- ***Terra preta* or biochar**: This ancient method revitalizes depleted soil fertility and carbon sequestration with fine-grained charcoal, rich in organic carbon.[21]
- **Grass-fed and grass-finished beef**: These cattle have nutritional and many other land health benefits with a much smaller carbon footprint of only one calorie of fossil fuel to produce two calories of food, compared to five to ten calories of fossil fuel for just one calorie of feedlot beef.
- **"Flerd" grazing**: The combination of flocks of sheep with herds of cattle, when properly managed, feeds carbon to fungi, protozoa, and nematodes, sequestering carbon and increasing plant species on one Australian ranch from 7 to 136 species.
- **Mob grazing**: Large numbers of cattle are lumped together in smaller areas for short time periods.
- **Strip-field farming**: Also known as an open field system, strip-field farming is a common agricultural and sustainable European system that was prevalent during the Middle Ages and lasted a thousand years. In this system families used long, narrow strips of land in the same field, utilizing banks of earth to divide fields and deflect wind for food crops.
- **Dormant-season grazing**: This practice allows cattle access to riparian and other sensitive areas for a short period in winter.
- **Terracing**: Converting slope land into a series of flat areas resembling steps can improve soil health and water availability.
- **Carbon-capture farms**: In areas such as the former wetlands on Twitchell Island east of San Francisco, Califor-

nia, research scientists with the U.S. Geological Survey in Sacramento have experimented with restoring subsiding wetlands, annually sequestering as much as twenty-five metric tons of carbon per acre.

- **Induced meandering**: This restoration and climate adaptation technique slows water, replenishes incised habitat, and sequesters carbon by stabilizing incised channels, nudging water back to its natural meandering flow with the installation of a variety of low-cost, bioengineered, grade-control structures.
- **Rooftop farms**: This expanding urban agricultural movement utilizes space on the rooftops of city buildings for both personal and communal climate gardens.
- **Organic no-till farming**: Farmers may elect not to utilize chemical fertilizers or disc-harrowing methods.
- **Agrivoltaic systems**: In these systems, the same land is used both for growing crops and for solar photovoltaic power generation.
- **Portable farms**: On-site food facilities can be used to raise fish (aquaculture, or fish farming) and plants (hydroponics, or growing plants in water).
- **Ocean farming**: Bren Smith, the founder of Green Wave, received the 2015 Buckminster Fuller Challenge prize for $100,000—one of the most prestigious sustainability prizes—for reimagining his storm-damaged farm as the world's first ocean farm. It's an underwater garden where seaweed is grown just below the surface, hanging vertically down next to mussels and scallops, and, below that, oysters.
- **In-forest biochar production**: Material is cut out of forests to reduce the fuel load and then processed with fire in a way to make carbon-rich biochar from slash.

These approaches are more frequently being used on farms, ranches, and other restoration or food production operations, as detailed in White and Pollan's *Grass, Soil, Hope*. While each of these practices offers hope, prompt and broad-scale application, or "scaling up," is the hurdle that awaits them.

When land is properly managed as part of these new economic models, tons of carbon per acre are removed from the atmosphere, improving the land's fertility, its capacity to hold water, and its ability to grow more and healthier food. Rangelands blanket approximately 770 million acres in the United States and constitute about one-third of the planet's ice-free land surface. Managing these ecosystems with an emphasis on carbon sequestration will remove a large portion of the global overheating, species extinction, and soil depletion threat. As we saw in chapter 6, each of us can take part in these solutions, no matter what size property we own or occupy.

There are those who believe a mere 2 percent increase in the organic content of the planet's soils, which consist of about 58 percent carbon, to 60 percent carbon could soak up all the excess carbon dioxide in the atmosphere in a decade.[22] In controlled experiments, the Marin Carbon Project applied compost and rotational grazing on a 540-acre cattle ranch in Northern California, sequestering about nine hundred extra pounds of carbon per acre. Christine Jones, founder of Amazing Carbon, thinks of healthy soils as liquid carbon pathways and is among the growing number of scientists who argue the restoration of ecological function in vegetation, soils, and waterways will result in a major drawdown of excess carbon dioxide from the atmosphere by way of bio-sequestration in soils.[23] These potential changes in the agricultural economy means the Paris Agreement's 350-ppm carbon dioxide target is achievable. James Hansen maintains that reaching the 350-ppm target "would require phasing out all fossil fuels by around 2030 and also drawing CO_2 out of the atmosphere and into the biosphere through massive afforestation, habitat restoration, and conversion to organic agriculture to rebuild carbon-sequestering living soils."[24]

The 2005 Millennium Ecosystem Assessment found we have already destroyed 60 percent of Earth's ability to support us.

Despite this bleak assessment, when habitat is managed to utilize nature's free gifts of sunlight, carbon, and subsurface life, the remaining 40 percent number can be increased. Replacing modern farming practices will result in carbon-rich topsoil, a benefit to farmers, rural communities, and the nation. Christine Jones emphasizes: "The key to subsurface management is above-ground management." Translation: abandoning the plow and adopting the alternatives listed earlier.[25]

Professor Rattan Lal believes the carbon in soil is like a cup of water, and we have removed more than half of it. However, like putting water back in the cup, good soil practices would replace the lost carbon and reverse global warming.[26] Lal maintains the adoption of no-till agriculture and other enlightened land-management practices could reduce carbon in the atmosphere on the order of 3 ppm per year. Although these estimates change often to reflect ever-worsening statistics and recalculations, at just a 3-ppm reduction a year, we could have the atmospheric carbon dioxide count back to the 350ppm level in less than twenty years.

Hopeful scientists and other land visionaries are no longer isolated outliers scrambling for viable solutions; they now constitute an encouraging, coalescing assortment of scientists, thinkers, writers, and activists who see a way out of the doomsday course of events that unfolded in spades with destructive weather in the last half of 2021 and throughout 2022 and 2023. Although they receive scant mention in the corporate-owned, fossil-fuel-aligned mainstream media, alternative news sources and the internet are helping fill the void. Dave Gori, former director of science for the New Mexico chapter of The Nature Conservancy, is one of these new breeds of thinkers. He has hope yet also makes clear that it's a matter of continuing to conduct on-the-ground adaptation experiments and of scaling up what is learned and applying it to the landscape.[27]

"Scaling up"—the prompt, broad-scale application of new adaptation tools—is the operative term in Gori's assessment and in the challenges facing soil loyalists. Biogeochemist Whendee

Silver of Silver Lab at the University of California, Berkeley—the lead scientist at the Marin Carbon Project—calculates that scattering half an inch of compost on half the grasslands in California could remove an amount of carbon from the atmosphere nearly equal to the amount emitted yearly from home and business electricity usage, could improve forage by 50 percent, and could retain 2,800 more gallons of water per acre. This addition would also increase the habitat's resistance to drought and improve its resiliency. Scientists tell us resiliency—"the ability to recover from or adjust easily to misfortune or change"—is one of the keys to "land health: the degree to which the integrity of the soil and ecological processes of rangeland are sustained over time."[28]

This hopeful assessment of the potential for a legacy-load drawdown of atmospheric carbon dioxide is only valid in a completely fossil-fuel-free economic model. In addition, scaling up new ranching and farming practices has its hurdles: transporting compost to rural areas is costly; steep or remote land is less suitable; and rangelands are not the same as what might be thought of as pasture land as addressed in the Marin Carbon Project. Many arid-land restoration enthusiasts warn against the "pastoral propaganda" handed out by those who see substituting cattle for buffalo and other pro-ranching ideas as cure-all restoration tools. Critics in the environmental community have issued warnings about the notion of cattle as a means of restoring land in the Southwest and caution that studies like the Marin Carbon Project create the risk of overgeneralizing the potential of carbon sequestration in arid habitats.[29]

Lucinda and I have cattle-ranching friends and neighbors who remain mired in denial about the global climate crisis. The American Farm Bureau once reported that 70 percent of farmers don't believe in human-induced climate change, although that number must surely be shrinking. Ignoring the science, they are unaware or reject the potential role the agricultural community can play as part of the solution to a problem that many of them insist doesn't exist. Agriculture has caused much of the problem yet has significant potential for helping curing it. Many farmers and ranchers also ignore the well-established truth that global overheating is hugely debilitating for agriculture.

Growing recognition of the link between the planet's warming and soil health can revolutionize the environmental movement and bring farmers, ranchers, and conservationists closer to a model pursued by collaborative groups like New Mexico's Quivira Coalition and Malpai Borderlands Group. Their model occupies the "radical center," a phrase coined by southern Arizona rancher Bill McDonald, and also a challenge found in the writing of Wallace Stegner—"the middle of the road is actually the most radical and most difficult"—used to describe an emerging and consensus-based approach to land-management challenges.[30] With approximately two-thirds of federally protected species on private land, a growing number of politicians and scores of others concerned with climate disruption recognize the world needs private landowner participation if we intend to arrest the warming of the planet.

There is a comforting hope in thinking soil will save us; recapturing carbon dioxide in soil with easy-to-execute natural climate solutions is a viable cure to the Trifecta Crisis. This soil-based drawdown of excess atmospheric carbon sounds simple because it is.

CHAPTER EIGHT

Strategies That Will Save Us

SCIENTISTS—AND NOW HORRENDOUS WILDFIRES and chaotic weather—have made clear that we must make prompt, radical changes to our lifestyles and behaviors, as well as structural changes in the economy and government, to remedy what is now feared by most scientists as irreversible. These fundamental changes in how to live need to occur at personal, national, and global levels. I'm hesitant to step too far beyond Lucinda's and my sphere of experience. We've learned how to restore habitat, sequester crises-causing carbon, and reap the benefits of natural climate solutions. Yet there is so much to be done; allow me to reach beyond restoration.

The reason for the etching at the head of this chapter is twofold. First, more than three decades ago, James Hansen (*right*) warned Congress about the climate crisis but, having been ignored, he's become an active member of the Voice of the Streets, protesting publicly and suffering arrest. John Wesley Powell (*left*), in the nineteenth century, also warned Congress of the flaws in its reckless settlement policy, yet he too was ignored. Powell endured an ignorant Congress and observed the suffering of thousands of pioneers, and if he was alive, he would be standing with Hansen. Second, the pairing of these scientists, the stupidity of Congress in ignoring them, and the risk of continuing to ignore Hansen and the now-unanimous scientific community means our experience will, by orders of magnitude, dwarf the tragedies endured by the newcomers drawn west by the mythic quarter section. Joining Hansen in the streets is among the most important actions we can take to solve these crises.

We're all challenged by the domineering, culturally engrained handicap that limits our potential to make a meaningful contribution to solving the Trifecta Crisis: *convenience*. The flawed lifestyle of convenience is not a matter of being lazy or not caring about the future; rather it's a multigenerational inheritance of a way of life, a deeply engrained culture. This addiction of ours is more insidious than a habit of mind. It's an inbred convention—obscured, but pervasive and governing. This is who we have become. It's believed by increasing numbers of scholars and citizens that consumption defines us. Convenience is the subliminal grease that keeps consumption central to our world.

The phrase "structural change" is tossed around a good deal, but the necessary revision to our way of life remains vague to many in terms of what we're being asked to change and how. The answer is just about everything: eating, vacations, air travel, clothing, entertainment, cars, and even lattes. Two hurdles impede our task of helping solve the Trifecta Crisis as individual consumers.

The first hindrance is the good fortune of many of us in the United States who fall into the top 1 percent of the world by income, making more than $34,000 per year. We don't see ourselves in such a rarified group, and its implications make little sense in terms of fault in how we've lived in this culture of enormous consumption. Our perspective is that the One Percenters are the Musks, Bezoses, and Gateses. Most of us don't have a well-informed understanding about how the climate-heating conduct of our everyday lives contributes to these crises. We don't see ourselves as super-emitters.

The second obstacle is one we share with the rest of the world—nobody takes responsibility. This deficiency is statistically well documented not just in the United States but throughout the world. The Kantar Group is a data research company based in London with a division called Kantar Public that promotes itself as working toward better public policy and a superior, fairer society. A 2021 poll carried out by Kantar Public in ten countries—including the United States, Britain, France, and Germany—and published in *Public Journal* found a significant gap between people's awareness and their action. When asked to indicate the three main environmental challenges, despite an encouraging 62 percent of people surveyed who acknowledged that the climate crisis was the main challenge facing the world—far more than pollution, waste impacts, or new diseases—only 36 percent rated themselves "highly committed" to preserving the planet and only 51 percent said they would definitely take individual climate action: "Citizens expect their national government to assume a large share of responsibility for protecting the environment, whilst considering themselves to be relatively good 'doing their bit' in this area. . . . While 78% of respondents say they would accept stricter environmental regulations, 45% say that they don't need to change their habits."[1]

Most troubling were the reasons given for their unwillingness to do more. Respondents gave the following reasons: there isn't agreement among experts on "the best solutions" (72 percent); "need more resources and equipment to act" (69 percent); "can't financially afford to make efforts needed" (60 percent); "missing information and clear guidance" on what to do (55 percent); "don't believe individual efforts can really have an impact" (39 percent); "there are reasons I should, but also reasons I shouldn't," suggesting environmental threats are overestimated (35 percent); and "uncertainty about their own capacity to change . . . don't really need to change their habits" (46 percent).[2] A survey of seven European countries by YouGov found the same thing: people are worried about climate change, happy with their government taking steps to counter it, but far less pleased with measures that would change their lifestyle, meaning: "I care, but don't inconvenience me."[3]

Americans personally committed to preserving the planet number well less than half at 40 percent, with only 21 percent believing their government is committed to preserving the environment and the planet. At least here in the United States, we're so accustomed to so many conveniences that asking us to change behavior we never think about is like asking us to change our pattern of breathing or to dream differently. What the Trifecta Crisis requires of us is so fundamental as to seem improbable. But we must bite the bullet and make the effort.

YOUR ACTION PLAN

Four of the best books on how we can save ourselves on national and global levels are *This Changes Everything: Capitalism vs. the Climate* by Naomi Klein; *Drawdown: The Most Comprehensive Plan Ever Proposed to Reverse Global Warming*, edited by Paul Hawken; *The Future We Choose: Surviving the Climate Crisis* by Christiana Figueres, executive secretary of the UN Framework Convention on Climate Change, and Tom Rivett-Carnac; and *All We Can Save: Truth, Courage, and Solutions for the Climate Crisis*, edited by Ayana Elizabeth Johnson and Katharine K. Wilkinson.

These books address the climate crisis more from a societal or governmental perspective, but we can also assume personal responsibility for our part in these crises. We can adopt a personal climate crisis strategy (see appendix C); urge city, county, and state governments to simply turn to Stanford University's Solutions Project (set out in appendix D and discussed later in this chapter); support Our Children's Trust litigation; donate to the Center for International Environmental Law or similar organizations; and most importantly, participate in the Voice of the Streets.

The focus here is on nature-based solutions and policies that each of us can adopt in our lives, as well as options that we can support outside of waiting on our government to do what it has so far failed to do. Beyond the universal "R rules" of reduce, recycle, refuse, reuse, repurpose, rethink, reclaim, and repair, there are a number of specific, actionable steps everyone can take immediately for achieving a reversal of humankind's assault on Earth. It's time we expand the planet-saving "Re-World" to include revegetate, restore, regenerate, regreen, rewild, reforest, and refurbish. This climate action agenda can be addressed selectively by each of us, but collectively, these items all need to be addressed by most of us. A straightforward way to initiate these life changes is to adopt three to five easy steps, then select another one each week until you've exhausted your capacity.

DEEP STRUCTURAL CHANGES

- **Honor science-based truth**. In her September 2019 remarks to the House Climate Crisis Committee and House Foreign Affairs Subcommittee, Greta Thunberg handed out a climate-crisis document prepared by the Intergovernmental Panel on Climate Change and said, "I am submitting this report as my testimony because I don't want you to listen to me, I want you to listen to the scientists. I want you to unite behind science."[4]
- **Promote, cultivate, and applaud virtue in public and private life**. Although virtue has lost much of its weight and meaning, our founders thought it was central to a respectable life and good government. Treat others respectfully, but when members of either your group or those opposite you violate fundamental principles, speak out against them. Refocus on the public good. Because the principle appears nearly absent in our society, permit me to summarize it: Civic virtue means individuals should have—and place above their own desires—a duty to their communities and their societies, a dedication to the common welfare even at the cost of their individual interests: self-sacrifice, participation, doing one's part, collaboration.
- **Abandon our convenient and consumptive way of life**. This may be the most difficult but most important adjustment we can make to save us from ourselves: making lifestyle and behavioral changes. Radical reduction in human consumption toward a sustainable economic model is imperative. The Transition Town Movement is one of many campaigns to initiate the shift from a fossil-fuel-based community to a sustainable community grounded in less consumption, local food production, co-ops, and a circular economy focused on resilience. Absent massive commitment to limit consumption, the existing order—corporate profit and growth driving public policy—of endless accumulation of goods and capital will persist.

- **Reduce the human population**. Have no more than two children, preferably fewer. Carter Dillard of the Fair Start Movement maintains the most significant way for an individual to reduce their carbon emissions is to have one fewer child, a decision that will reduce carbon dioxide emissions to more than fifty-eight tons for each year of a parent's life. In the United States, the carbon legacy and greenhouse gas impact of an extra child is almost twenty times more important than practices like driving a low-mileage car or recycling.
- **Rethink land use**. There are twenty-one ways—natural climate solutions—to use land differently that "can provide up to 37% of cost-effective CO_2 mitigation [and] also offer water filtration, flood buffering, soil health, biodiversity habitat, and climate resilience."[5] The options range from biochar to better forest management, reforestation, no-disc farming, and wetlands restoration, to name a few.

PUBLIC INVOLVEMENT

- **Never miss an opportunity to vote**. It's a terrifying realization that 98 percent of incumbents are reelected and that statistically they ignore the masses and coddle wealthy donors. When 71 percent of voters favor higher taxes on those who earn more than $1 million a year from capital gains but politicians are reluctant to act, something is wrong. Vote out every incumbent who demonstrates any hesitation about the severity of these crises. Show up at their speaking events and tell them they'll not have your vote without their commitment, right then, to raise taxes on the rich and address the climate, species extinction, and soil loss crises.
- **Become an activist**. Engage in the stand-up, nonviolent, but aggressive and full-throated variety of activism. The social contagion that flows from the fundamental Voice of the Streets is the future. Schedule a face-to-face appoint-

ment with your city manager or city council to adopt the Stanford University Solutions Project. (Appendix D is a draft memorandum about the Solutions Project that you can modify to your liking.) Become a consumer activist too. Telephone and send follow-up letters to companies to explain how their packaging can become environmentally friendly or to encourage recycling.

- **Lobby your corporate employer or university to divest.** Employees of corporations and big companies with retirement plans and students at universities can form groups to demand that these organizations divest themselves from the fossil-fuel industry.

DAY-TO-DAY CHANGES

- **Limit airline travel.** Eleven percent of greenhouse gas emissions are attributed to air traffic and transport. We need to eliminate virtually all air travel until it transitions off fossil fuels. Take no more than one short-haul flight every three years and one long-haul flight every eight years. Telecommunication is the key to eliminating physical travel. Vacation via air never. Plan additional time to travel by train.
- **Eliminate most meat and dairy products from your diet.** Eat meat only once a week; eat vegetarian as often as possible. Investigate alternatives to animal proteins and dairy products.
- **Get rid of your personal motor vehicle.** The transportation sector contributes a third of U.S. carbon emissions. If you can't get rid of your car, hold on to it as long as you can. If you must keep your car, limit your automobile use by carpooling, biking, or walking. Also, modify your driving habits: never drive above the speed limit, and check your tires for proper inflation to save gas.
- **Divest your portfolio of fossil fuel investments.** Cities, universities, churches, and other institutions have already

begun this process, but it needs to be universal and personal. Without investors, polluters perish. If you are fortunate enough to invest in the stock market, make sure your investments are in socially responsible enterprises. Fossil fuels, never.

- **Prioritize zero-carbon sources of renewables**. In New Mexico, the oil and gas industry is responsible for 53 percent of greenhouse gas emissions.[6] The most comprehensive approach to combating the climate crisis is to abandon our dependence on fossil fuels, pull out the massive military presence—the United States has nearly eight hundred bases around the world and recently added four in the Philippines—required to maintain access to this so-called cheap energy from the Middle East, and use the savings from reducing that bloated and expensive military for a Great Transition to clean, renewable energy sources. A military pullout is obviously a governmental function, but we all can urge our representatives to eliminate this waste.
- **Build grade-control structures**. Either on your own land or with a habitat restoration group, install one-rock dams, other types of erosion-control structures, and other natural climate solutions. There are hundreds of habitat restoration organizations in need of volunteers.
- **Use less water**. Switch to low-flow fixtures; shower instead of bathing; avoid overuse of running water; and Google how to use less water.
- **Restore soil and the organisms living there**. Restore soil, knowing it will help save us if we nurture it and its inhabitants to health and well-being.
- **Become part of the post-waste world**. Reduce, recycle, refuse, reuse, repurpose, rethink, reclaim, and repair, and revegetate, restore, regenerate, regreen, rewild, reforest, and refurbish.
- **Replace your gas stove**. The methane emissions from gas stoves in the United States are equal to adding half million cars to the highway yearly. One in eight cases of childhood

asthma is linked to gas stoves, which emit nitric oxide, nitrogen dioxide, carbon monoxide, particulate matter and formaldehyde, benzene, xylene, toluene, and ethylbenzene.

- **Weatherize wherever possible**. In the United States, buildings use one-third of all energy consumed and two-thirds of all electricity. A full one-third of carbon dioxide pollution is caused by heating and cooling homes in need of retrofitting and their appliances modified or replaced. Improving the energy performance of your house or business will reduce energy consumption, save money, and create green jobs. The American Council for an Energy Efficient Economy maintains that intelligent efficiency, smart building, and smart manufacturing are a critical part of the crisis solution.
- **Create a climate garden**. Anyone with a yard can plant a garden, and those without land can work in a community garden.
- **Abandon fashion**. According to the United Nations Environment Program, the fashion industry is responsible for 10 percent of annual global carbon emissions. Limit the purchase of new clothes to no more than three new items per year. Donate infrequently used items, and don't replace them. Other than for growing children, buy far less clothing.
- **Mend clothing**. Extend the life of clothing: "make do and mend." Take used items to a secondhand store.
- **Plant tree seedlings**. Forest restoration can have more potential than climate gardens. It "isn't just one of our climate change solutions, it is overwhelmingly the top one," says Professor Tom Crowther at the Swiss university ETH Zurich. Crowther led "The Global Tree Restoration Potential," research published in the journal *Science* that maintains a tree planting campaign—there are roughly 64,100 species on the planet—to increase tree population globally can sequester two-thirds of the roughly 300 billion metric tons of carbon that humans have put in the atmo-

sphere since the dawn of the Industrial Revolution. This means growing 1.2 trillion new trees and 50 to 100 years of carbon dioxide drawdown. Humans have felled 46 percent of all trees since the dawn of civilization. The point is, says Crowther: "reforestation is so much more vastly powerful than anyone expected. . . . What blows my mind is the scale. I thought restoration would be in the top 10, but it is overwhelmingly more powerful than all of the other climate change solutions proposed. . . . [Tree planting] is available now, it is the cheapest one possible and every one of us can get involved."[7] In addition to carbon capture, due to the way trees physically transform energy and water also keeps air cool and moist, at least half a degree Celsius, even at great distances. As with climate gardens, each one of us can make a tangible impact by growing trees, donating to forest restoration organizations, and boycotting irresponsible companies that fell trees in their production processes.

- **Replace or shrink your lawn**. Revegetate your smaller lawn with a well-chosen, self-sustaining native plant community. Plants eat sunlight to feed insect herbivores eaten by birds. Yet introduced plants such as lawn grass provide little of the food for animals that evolved with native plants. It's not the number of flora and fauna species in a habitat that determines ecosystem function but the interaction diversity between them; it is "the way species interact with one another, and not the actual species themselves, that forms the glue holding nature together."[8] More than a gigaton of carbon would be drawn down from the atmosphere over two decades if a third of the world's lawns were replaced with trees.
- **Eliminate introduced and invasive plants**. Identify the plants on your property and where they are from. If they are not local, they're not serving local animals with food.
- **Avoid single-use products**. Eliminate all single-use products from your daily life, excepting medical single-use

items. There are 265 billion disposable paper cups thrown away yearly. By refusing use, we can force manufacturers and retailers to assume their share of responsibility. Take a glass container to pick up take-out pizza and coffee.

- **Eliminate plastics, petrochemicals, petroleum, other fossil fuels and chemicals**. Plastics are easier to avoid because you can see them. The world consumes more than a million plastic bags a minute. Petrochemicals are unseen, the two most common being olefins and aromatics used in solvents, detergents, resins, lubricants, gels, and fibers. To eliminate consumption of petrochemicals, be sure to read labels intensely, and do your research on what you buy. "Forever chemicals," toxic per- and polyfluorinated substances, were detected in every one of thirty thousand umbilical cord blood samples across forty studies conducted from 2018 to 2022.[9]
- **Counsel children**. It's not only adults who are concerned about the Trifecta Crisis. Children realize they are the generation at risk. We need to learn about these crises and their solutions and how we can solve them and teach our children how to live responsibly. Children need hope. Children are forcing adults to change.
- **Grocery shop differently**. Shop in bulk and bring containers and bags instead of using store-supplied disposable ones. If you want a mantra to help prevent civilizational death, what some are calling a "collective suicide pact," it's "cars, planes, and meat."

Our individual actions can amount to a mountain-moving force once enough of us sign up, adopt climate crisis ambition, and do our part. This is not a minor ask. Few of us have made the necessary changes, adjustments, and adaptations on any scale near what is needed. It requires us to abandon a hard-to-identify array of culturally inherited conveniences that we'll struggle to even recognize, much less forgo altogether. This change requires all of us to take stock of what it means to be moral people. I was grocery shopping with my sister when the cashier asked her, "Paper or plastic?" She answered, "Doesn't matter." And I had this exchange with a colleague working with Th!rd Act Silver City, where I suggested we forward to members an email from Our Children's Trust asking us to contact Attorney General Merrick Garland and urge him to end the U.S. government's seven-year scorched-earth litigation efforts to avoid the *Juliana v. United States* litigation. Smart, educated, and progressive, my colleague responded: "No idea what this is about." I explained the nonprofit group's litigation efforts to address the cause of the climate crisis and where to look for details; he did, then responded: "I'm sure this is a fine undertaking, but I'm happy to wait until it comes to some sort of fruition, or doesn't. I looked at the Wikipedia page. While wholly laudable, it seems entirely likely this won't really go anywhere. But if it does, I'm prepared to pay attention to it." Borrowing a thought from Martin Luther King Jr., the greatest threat to solving the climate and companion crises is not denialists; rather, it's the indifference of otherwise good people.

In the context of the Trifecta Crisis, to determine what it means to be that moral person requires us to step back and do some accounting about where we are and how we got here. I have a thoughtful friend who startled me with what he understood as the Darwinian notion that the current state of affairs on the planet is about what should be expected from a biologically superior and inordinately successful species. Overrun by eight billion humans, our planet has been overharvested, overdrilled, over-pumped, over-mined, and polluted in land, air, and water; thousands of species have gone extinct. His thinking, simplified, "was pure Darwinian, obedient to the gods of unlimited growth and reproduction."[10] This thinking parallels the brown tree snake's successfully overtaking all species but humans on Guam. From our position as the highest ranking, dominant species on the planet, my friend asserts, humankind is headed right where Darwin says we would go. Despite the dire outcome, he's satisfied that we're predictably headed to the edge of a small, polluted, nutrition-less world where life perishes.

In the 1920s, a twenty-four-year-old Russian student by the name of Georgy Gause—best known for his competitive exclusion principle—put forward the idea that no two species with similar ecological niches could coexist in a stable equilibrium and that one would out-produce the other to extinction. In one of his experiments, Gause boiled a half grain of oatmeal in water and poured the broth into a test tube with one single-celled protozoon. Repeating the experiment one hundred times, he created one hundred micro-earths with a limited food source and a creature with the same purpose as all life-forms: to produce more of themselves toward ensuring their biological future.

At first the protozoa reproduced slowly, but after a time their numbers spiked in a frenzy of exponential growth until their progeny ran out of nutrients near the edge of the petri dish. With the food exhausted, life dwindled, and they either drowned in their waste or starved from lack of food. Our case is different in scale but not principle; there is reason to fear the fate of humankind parallels the occupants of Gause's petri dish. Bacteria in a petri dish are similar to other life-forms: once established, they will reproduce at a remarkable rate until they reach the limits of their surroundings and eventually perish. Constraining one's own growth defies biological imperative. Burmese pythons in Florida; water hyacinths in African rivers; the ʻaeʻae wetland plant in Hawaiʻi; rabbits in Australia; Asian long-horned beetles, gypsy moths, and Zebra mussels in the United States; and brown tree snakes in Guam are all imported, invasive species that have no natural predators in their new homes and therefore are overrunning environments without regard to anything but their singular biological imperative, racing toward the edge of the metaphorical petri dish.

While my friend's interpretation of Darwin is consistent with the petri dish experiment, I would not have expected his or anyone's approval for the endgame of such an intelligent species to be self-extinction. Aren't we deserving of more credit that the brown tree snake? As one of millions of Earth's life-forms, humankind has prospered, persisted, prevailed, and established itself as the most accomplished aboveground species ever. Part of the explanation for our success is something absent in my friend's thought: meaning, that is, what it means to be human. His line of thinking fails to consider fundamental questions about purpose or values or virtue. To him, those are an entirely different matter, questions of morality or philosophy, not matters that concern the biological world of humans and other animals.

A biological drive toward extinction strikes me as derivative of the so-called selfish-gene revolution introduced by American evolutionary biologist George C. Williams in his 1966 book, *Adaptation and Natural Selection*. Popularized by Richard Dawkins's 1976 book, *The Selfish Gene*, this "revolution" espouses the powerful notion of self-interest. Ayn Rand novelized the absolute of biological self-interest in her 1957 *Atlas Shrugged*, in which protagonist John Galt speaks for her view that human nature is fundamentally selfish: "By the grace of reality and the nature of life, man—every man—is an end in himself, he exists for his own sake, and the achievement of his own happiness is his highest moral purpose."[11] Rand's philosophy appears to be alive and well as the dominating ethos of the super rich.

Although this thinking is common today, it's a modestly evolved line of low-effort thought that ignores both virtue and the robust scientific evidence supporting altruism as a fundamental human trait and one existing in other animals. It's well-documented wildlife behavior: a leopard caring for a baby baboon, dolphins circling human swimmers to protect them from sharks, and sperm whales caring for an injured dolphin. Although I take exception with altruism as a suitable foundation for a land ethic in an era of climate, species, and soil crises (chapter 9), altruism is an admirable trait with a deep history in life. Charles Darwin understood the rational and evolutionary implications of human altruism, as he set out in his 1896 *The Descent of Man*:

> It must not be forgotten that although a high standard of morality gives but a slight or no advantage to each individual man or his children over the other men of his tribe, yet that an advancement in the standard of morality and an increase in the number of well-endowed men will certainly give an immense advantage to one tribe over another. A tribe including many members who, from possessing in a high degree the spirit of patriotism, fidelity, obedience, courage, and sympathy, were always ready to give aid to each other and to sacrifice themselves for the common good, would be victorious over most other tribes; and this would be natural selection. At all times throughout the world tribes have supplanted other tribes; the standard of morality and the number of well-endowed men will thus everywhere tend to rise and increase.[12]

This idea is simple enough: altruistic groups will prevail over comparably sized non-altruistic groups. Altruism is the biological trait lying at the core of civic virtue.

In contrast to the legions of vocal Rand adherents in today's American political climate, Matt Ridley's 1996 book, *The Origins of Virtue: Human Instincts and the Evolution of Cooperation*, builds on Darwin's thinking and debunks the notion of "man as an end in himself" or humankind consuming all of life and ultimately extinguishing itself in the global petri dish.

> The virtues of tolerance, compassion, and justice are not policies towards which we strive, knowing the difficulties along the way, but commitments we make and expect from others. . . . The virtuous are virtuous for no other reason than that it enables them to join forces with others who are virtuous, to mutual benefit. And once cooperators segregate themselves off from the rest of society a wholly new force of evolution can come into play: one that pits groups against each other, rather than individuals.[13]

Altruism is now accepted as a fundamental trait in nature. Among many of the most accomplished thinkers, it is deeply embedded in E. O. Wilson's thinking for both humans and nonhumans: "Selection among genetically diverse individual members promotes selfish behavior. On the other hand, selection between groups of humans typically promotes altruism among members of the colony. Cheaters may win within the colony, variously acquiring a larger share of the resources, avoiding dangerous risks, or breaking rules; but colonies of cheaters lose to colonies of cooperators."[14] The adjustments proposed here to our lifeway are solutions we can pursue individually, yet it's clear that survival is a collaborative process that we need to pursue—not as cheaters but as colonies of cooperators. In addition to what we can accomplish on our own by making personal changes to our way of living as cooperators in a colony of like-minded people who prioritize future generations, we can also lend support to nongovernmental entities whose goal is to solve the Trifecta Crisis. There are many, but Lucinda and I follow the progress and provide financial support to three in particular.

THE SOLUTIONS PROJECT

Mark Jacobson, civil and environmental engineering professor at Stanford University, and his colleagues have a plan to begin transforming the U.S. from dependence on fossil fuels to 100 percent renewable energy by 2050. The fifty-state plan is posted on the website of Stanford's Solutions Project, a nonprofit outreach effort run by Jacobson's group to raise public awareness about switching to clean energy produced by wind, water, and sunlight. It includes an interactive map with an explanation of how each state can meet virtually all of its power demands (transportation, electricity, heating, etc.) no later than 2050 by switching to a clean technology portfolio that is 55 percent solar, 35 percent wind, 5 percent geothermal, and 4 percent hydroelectric. Nuclear power, ethanol, and other biofuels are not included in the proposed energy mix for any state. Energy behemoths are excluded from the plan as unnecessary. A study by the 2020 energy-policy group Rewiring America concluded an aggressive push toward 100 percent renewable energy would save American households as much as $321 billion, or up to $2,500 per household per year.

The Solutions Project's carbon dioxide abatement plan is but one of a number of recommendations for making the transition of billions of dollars in annual subsidies from public coffers to renewable energy. A February 2022 study by the B Team and Business for Nature found that $1.8 trillion in annual subsidies—to the fossil fuel industry ($620 billion), agriculture ($520 billion), water ($320 billion), and forestry ($155 billion)—are leading to our own extinction when these funds could be repurposed toward survival.[15] It's entirely possible for wealthy countries and regions to simply end these subsidies and abandon fossil fuels altogether. They could completely wash their hands of the high-consumption, carbon-hungry, greenhouse-gas-spewing system. The science is unequivocal that countries can develop renewable energy infrastructures within a twenty- to forty-year timeframe rather than continuing on their current paths of doling nearly $2 trillion in subsidies to these polluting industries and condoning their treatment of the atmosphere as a waste dump. To be precise, the $620 billion number for the fossil fuel industry increases to more than $5 trillion when you take externalities (pollution and other damage to the planet) into account.[16]

Among the earliest studies to estimate the future economic impact of meeting the 2015 Paris Agreement goals, "Large Potential Reduction in Economic Damages Under UN Mitigation Targets," published in the journal *Nature*, found most nations, representing 90 percent of the global population, would benefit economically if we kept the globe from warming no more than 1.5 degrees Celsius above preindustrial levels. The cost? $5 trillion. The savings? About $30 trillion. A 2021 study published by the World Resources Institute, New Climate Economy, and the International Trade Union Confederation concludes, for every $1 million the U.S. government invests in solar energy, the investment produces 1.5 times more jobs than exist in fossil-fuel-based energy; investing in building efficiency spurs 2.8 times as many jobs; and investing in ecosystem services, such as the restoration we do here, creates 3.7 times as many jobs as do investments in oil, gas, and coal, further making the case that a green economy is a far more labor-intensive economy.[17]

There's a little-recognized financial barrier ahead in the transition to a fossil-fuel-free future. *Stranded assets* is a term for a company's investment rendered unproductive because government regulations—such as a dramatic change in environmental policy—or other factors results in the loss of their projected value. The *Nature Climate Change* study "Macroeconomic Impact of Stranded Fossil Fuel Assets" suggests there is a "climate bubble" due to the overvaluing of fossil fuel assets that will end up being stranded assets as a result of new advances in technologies for energy efficiency and renewable power. The accompanying drop in cost makes low-carbon energy much more economically and technically attractive. The authors maintain the Paris Agreement's aim to limit the increase in global average temperature to well below two degrees Celsius above preindustrial levels requires all but a fraction of existing reserves of fossil fuels and production capacity remain unused, hence becoming stranded assets. Because investors assume these reserves will be commercialized, the stock prices of listed fossil fuel companies appear to be overvalued, giving rise to this underemphasized carbon bubble. The financial implications of the requirement to "keep it in the ground" possess one of the major hurdles to solving these crises. There is as much as five times more fossil fuel on the books than climate scientists warn is safe to burn, and 80 percent of those reserves must remain in the ground, forever unused, if the climate crisis disaster is to be avoided.

The obvious problem with the civilization-preserving requirement of radically limiting carbon dioxide by leaving 80 percent of reserves in the ground is that these reserves and hundreds of billions of dollars in capital investment would become stranded and the corporate stock of Exxon, Shell, BP, and the other fossil fuel companies and oil-producing countries would plummet. When you convert the 80 percent fossil fuel "hold-back" requirement to stock value (and the cost of stranded assets), investors' holdings are worth not much more than twenty cents on the dollar, maybe less.

John Fullerton, a former managing director at JPMorgan, "calculates that at today's market value, those 2,795 gigatons of carbon emissions [contained in current fossil fuel reserves] are worth about $27 trillion," meaning an 80 percent write-off will cost investors almost $22 trillion.[18] Enter money, politics, and economics. One of several reasons why climate disruption has posed such a conundrum is because this issue implicates far more than just science. There are economic and political market devotees who believe the invisible hand of the free market can address the problem. But this infusion of free-market fundamentalism into the scientific equation of climate disruption is a civilization-threatening red herring. Even Nicholas Stern, former chief economist of the World Bank, president of the British Academy, and currently chair of the Grantham Research Institute on Climate Change and the Environment at the London School of Economics, has recognized the free market alone is inadequate and that global warming is "the greatest example of market failure we have ever seen."[19]

The feasibility of the transition to renewable energy is old news to those who have been studying the potential of the transition and to others paying attention. While Stanford's Jacobson is dismissive of the climate change doubters, he remains realistic about the challenge—the social and political will—to achieve transformative, wide-ranging, cultural developments to end fossil fuel addiction and move to weather-driven renewable energies:

> It's absolutely not true that we need natural gas, coal and oil—We think it's a myth. . . . [The transition from fossil fuels to renewables] . . . involves a large-scale transformation. . . . It would require an effort comparable to the Apollo moon project or constructing the interstate highway system. But it is possible, without even having to go to new technologies. We really need to just decide collectively that this is the direction we want to head as a society.[20]

Fossil fuel madness needs to stop. It's simple. As they say, this is not rocket science:

- We have eliminated one-half of Earth's topsoil in the last 150 years.
- In one example of the Trifecta Crisis, we have created a junk pile of plastic fishing trash in the ocean between Hawai'i and California that is twice the size of the state of Texas.
- We have increased atmospheric carbon dioxide more than 144 ppm since the Industrial Revolution. (Around 424 ppm is about where we are in 2023; it was less than 280 ppm in the 1800s.)

Industry loyalists squawk "simplistic." Among their political talking points is the claim that it's unrealistic to go green so soon. Yet the transition to renewables is not complicated; governors in all fifty states need to simply pick up the phone and tell their chiefs of staff to go to the Solutions Project web page and see to it that the recommendations for their state are implemented, posthaste. The oil-industry-fueled doubt over the transition out of fossil fuels is fraudulent. According to the 2018 *State of the Carbon Cycle Report* issued by the U.S. Global Change Research Program, the cost of modifying our energy infrastructure and reducing emissions by 80 percent below 2005 levels by the middle of the twenty-first century—a goal consistent with the UN's two-degree target—would only cost the United States between $1 trillion and $4 trillion (in

2005 U.S. dollar values, spread over more than three decades) by midcentury. This cost, plus the 3 to 5 percent loss in gross domestic product, would be offset by savings related to climate-related damages forecast at $170 to $216 billion per year by 2050 if we proceed with business as usual. The initial step to creating a pollution-free world and alleviating climate disruption could begin with those governors' phone calls. Congress would follow, and next, the world. This is a decision the global collective needs to make, and the United States can lead. With upward of 85 percent of the world's energy derived from the burning of fossil fuels and much of it produced by the top 10 to 20 percent of the world's people by income level, the "haves" need to lead. The Voice of the Streets is the only way to force those in power to do what they must. We must never forget that every person's fossil fuel use anywhere affects everyone everywhere, and every emitted carbon molecule counts.

The 2021 IPCC report left us with "hesitant hope"; assuming we take drastic action now, we'll still only avoid the worst. We have already passed four of the nine planetary boundaries for survival: extinction rate, deforestation, atmospheric carbon level, and flow of nitrogen and phosphorous into the oceans. Using terms like "extraordinary," "terrifying," and "uncharted territory," scientists report a series of three days in July 2023 were the hottest worldwide since records have been kept, probably as far back as 100,000 years. Humankind has never lived on the planet this hot.

OUR CHILDREN'S TRUST

Cities, states, and counties are initiating climate impact litigation. Our Children's Trust and similar groups have initiated a spate of litigation in the United States, the trust's claim brought on behalf of children, challenging the federal and state governments for violating the children's constitutional rights to life, liberty, property, and equal protection by failing to responsibly address climate crisis–related harms. Should the trust prevail, what Thomas Kuhn called a paradigm shift will follow; it "would necessitate that the government realign its subsidy permitting regulation around the energy system (obviously systemic) to meet the requirements of the court's decision, however that may be written. The areas where the US could then choose to adjust its policies to bring themselves into constitutional compliance are system-wide: transportation, agriculture, infrastructure, economic, energy, national security, etc."[21]

Although there is much to be encouraged about, it must be acknowledged the government under Presidents Barack Obama, Donald Trump, and now Joe Biden have gone to extraordinary lengths to avoid trial. The central piece of Our Children's Trust's U.S. litigation—*Juliana v. United States*—has struggled against the government's successful pretrial appeals yet still holds promise. The case has been pending since Obama's presidency and has suffered a myriad of scorched-earth, enormously unusual pretrial government appeals. The initial rulings were encouraging. A groundbreaking April 2016 ruling by the Federal District Court magistrate in Eugene, Oregon, Judge Thomas Coffin, rejected a request by the government and fossil fuel industry interests to dismiss the litigation. The court reached this clear-eyed conclusion: "If the allegations in the complaint are to be believed, the failure to regulate the emissions has resulted in a danger of constitutional proportions to the public health."[22] Although the magistrate noted that the lawsuit was "relatively unprecedented," he agreed that the children have standing to sue and the dangers, as alleged, are comprehensive:

> disintegration of both the West and East Antarctic ice sheets with concomitant sea level rise damaging coastal regions; changing rainfall and atmospheric conditions affecting water and heat distribution causing severe storm surges, floods, hurricanes, droughts, insect infestation, reduced crop yields, increased invasive vegetation, and fires; ocean acidification

damaging sea life; increase in allergies, asthma, cancer, and other diseases; and harm to national security causing destabilization in various regions of the world.[23]

Based on the public trust doctrine, the suit was filed on behalf of twenty-one young people, aged eight to nineteen. James Hansen, who brought global warming to the attention of Congress in 1988, is participating as a guardian for plaintiff "future generations" and his own granddaughter. Alleging concrete harm from carbon emissions and seeking to force the government to commit to significantly and swiftly reducing carbon dioxide emissions and to adopt a science-based recovery plan, this litigation zeroes in on the essentials of our crises. The executive director of Our Children's Trust, Julia Olson, said the initial decision was remarkable: "There's really not been anything like it yet. . . . It will be the first time that the federal government's fossil fuel policies will really be looked at in accordance with the Constitution and their obligation to protect young people and future generations."[24] Bill McKibben and Naomi Klein called the case the "most important lawsuit on the planet right now." Our Children's Trust's attorney Phillip Gregory added, "This decision is one of the most significant in our nation's history."[25] The lawsuit could redeem the promise of the "self-evident, unalienable rights" of "life, liberty, and the pursuit of happiness" rights assured in the Declaration of Independence.

After a long period of scientific investigation, Stanford Research Institute scientists submitted a report to the American Petroleum Institute in 1968 advising the fossil-fuel industry carbon dioxide was the only pollutant that had been proven to be of global importance to the environment and that carbon emissions would cause serious worldwide environmental changes. When Olson was asked about it, she replied,

> That issue is central to the case. . . . We have evidence that since 1965, and even the late '50s, the federal government has

> known that by continuing to extract and burn fossil fuels, that it would cause catastrophic consequences for future generations. . . . The information of the fossil fuel industry actually is *predated* by knowledge by our federal government. In collusion, both the fossil fuel industry and our federal government have been knowingly causing climate change through their acts. . . . The collusion, the destruction of the climate system—is what we will present in trial in Oregon.[26]

The magistrate's ruling was momentous and offered hope for those litigating the risks of climate overheating and challenging the government's indifference to these crises. When the ruling was forwarded to U.S. District Court Judge Ann Aiken, she accepted the magistrate's decision with the following approval: "Exercising my 'reasoned judgment,' I have no doubt that the right to claim a climate system capable of sustaining human life is fundamental to a free and ordered society."[27] With a bold pen, Judge Aiken said the climate crisis needs to be addressed and expressed her thinking in terms of "all deliberate speed," a potent phrase in U.S. jurisprudence, language derivative of the 1955 Supreme Court's momentous decision in *Brown v. Board of Education*'s ordering of the end of racial segregation in schools. The excruciating details of the years of back-and-forth between Judge Aiken's trial court, the Ninth Circuit Court of Appeals, and the U.S. Supreme Court is beyond the scope of this book, but as of this writing, despite breaking new ground, the litigation has been stalled. Ensnared in the government's turbulent appellant waters, the case had been gasping for air. Yet it's now, as of July 2023, that Judge Aiken has granted the trust's motion to amend its complaint, putting the case back on track toward trial in order to cure shortcomings said by the appellate court to forestall the case's going forward.

Arguably one of the more promising of a number Our Children's Trust cases is the August 4, 2021, ruling in the Montana litigation of *Held v. State of Montana* in which District Court Judge Kathy Seeley denied the state's motion to dismiss in a decision based in part on article II, section 3, and article IV, section 1, of the Montana constitution, which provide that "all persons . . . have certain inalienable rights. They include the right to a clean and healthful environment" and "the State and each person shall maintain and improve a clean and healthful environment in Montana for present and future generations." Additionally, under article IV, section 1, subsection 3, "The legislature shall provide adequate remedies for the protection of the environmental life support system for degradation and provide adequate remedies to prevent unreasonable depletion and degradation of natural resources."[28] The first-ever constitutional climate trial in the United States concluded June 20, 2023. If Our Children's Trust prevails in *Held*, it will serve as a foundation for the group's companion cases in other jurisdictions, especially New York, Pennsylvania, and other states that have similar constitutional guarantees to a healthy environment. The judge asked to see proof of Our Children's Trust's claim that the executive and legislative branches of government aggressively and recklessly perpetuated a fossil-fuel-based energy system, knowing all along the existential dangers their conduct would cause.

Along with the *Juliana* and *Held* cases, Our Children's Trust has initiated litigation on behalf of children in all fifty states and, as of this writing, litigation is pending in Alaska, Florida, North Carolina, Oregon, Colorado, Washington, Virginia, Utah, and Hawai'i. Internationally, sixteen children, including Greta Thunberg, have initiated a complaint with the United Nation's Convention on the Rights of the Child against Argentina, Brazil, France, Germany, and Turkey, asserting that these countries have breached their treaty obligations to protect them from direct and foreseeable risk to their health and well-being posed by the climate crisis. Climate litigation is pending in at least 37 countries against governments and companies, 1,587 cases overall, with 1,213 in the United States, as of this writing. In a historic vote, on July 28, 2022, the United Nations General Assembly of 161 countries voted to recognize the right to a clean healthy, and sustainable environment.

In Mexico, in *Jóvenes v. Gobierno de México*, the Seventh Collegiate Court reversed a lower court ruling against the youth plaintiffs and ordered the court to hear evidence on the facts of the case. *La Rose v. Her Majesty the Queen* is pending in Canada. In yet another case, the Australian government recently took a drubbing:

> It is difficult to characterize in a single phrase the devastation that the plausible evidence presented in this proceeding forecast for the Children. As Australian adults know their country; Australia will be lost and the World as we know it gone as well. The physical environment will be harsher, far more extreme and devastatingly brutal when angry. As for the human experience—quality of life, opportunity to partake in nature's treasures, the capacity to grow and prosper—all will be greatly diminished. Lives will be cut short. Trauma will be far more common and good health harder to hold and maintain. None of this will be the fault of nature itself. It will largely be inflicted by the inaction of this generation of adults, in what might fairly be described as the greatest inter-generational injustice ever inflicted on one generation of humans upon the next.[29]

A number of these youth-based cases have been dismissed and ruled against the children based on one or more legal theories of avoidance: courts cannot solve political questions; states do not have a fiduciary duty to affirmatively protect the state's public trust resources from the effects of climate change; students don't have standing; or a plethora of technical rulings that allow the courts to avoid addressing the merits of these claims. It's of interest that appeals in many of these cases, such as Oregon's *Chernaik v. Brown*, are supported by amicus curiae (friend of the court) briefs.[30] The appeal in *Chernaik* was joined by more than 175 different individuals and organizations representing a wide variety of interests on behalf of hundreds of thousands of citizens who live, work, and recreate in the state and depend on public trust resources for their well-being. The Children's Trust lost the case.

The last Wednesday in May 2021 has been dubbed by some as Black Wednesday for the hydrocarbon oligarchy Big Oil and a turning point in the effort to force international oil companies to cut back greenhouse gas pollutants and to establish a strategy for a low-carbon future. Royal Dutch Shell was ordered by a court in The Hague to cut its emissions by 45 percent within the next ten years; ExxonMobil's rebellious shareholders voted to replace three of the oil giant's board members in favor of climate-sensitive candidates; and 61 percent of Chevron's investors voted in favor of a climate resolution forcing the company to reduce its emissions. Students from Yale, MIT, Princeton, Stanford, and Vanderbilt have asked the attorneys general in their respective states to investigate the schools' breaches of the Uniform Prudent Management of Institutional Funds Act by investing in the fossil fuel companies causing the climate crisis. The act requires university boards of trustees consider charitable purpose in investments, a responsibility to invest with prudence, and a duty to invest with loyalty. Harvard and Cornell have already announced they will divest themselves of fossil fuel investments in response to similar claims.

After I submitted this manuscript for publication, the flood of climate breakdown litigation throughout the world caused me to remove material that would be dated the first day the book became available.[31] Then this happened: sixteen municipalities in Puerto Rico upped the litigation assault for addressing the climate crisis with a first-ever lawsuit filed in U.S. Federal District Court in December 2022 under the 1970 Racketeer Influenced and Corrupt Organizations Act (RICO), a law created to combat criminal enterprises like the mafia. Although the RICO tactic is new, the underlying facts reflect the same collaboration alleged in other lawsuits between international oil and coal companies, trade associations, a network of think tanks, scientists and other intertwined colluders, creating a consortium—the Global Climate Coalition, with its formal action plan—committed to deceiving the public about carbon and the climate by telling us something they knew to be false. This litigation has been significantly bolstered by yet another 2023 study finding Exxon scientists not only had long known of the dangers of global heating but, as early as the 1970s, had accurately predicted the upward trend of global temperatures and carbon dioxide emissions that correspond with what the world has actually endured. It turns out that Exxon knew of its crime even before the births of many of the scientists now studying the crisis. Next, in April 2023, the U.S. Supreme Court rejected five appeals by ExxonMobil and other oil giants to remove climate lawsuits from state to federal courts, breaking a logjam that will lead to increasingly common city- and state-initiated climate litigation. Reaching an even higher potential litigation plateau than RICO, a 2023 law review article accepted for publication in the *Harvard Environmental Law Review*—"Climate Homicide: Prosecuting Big Oil For Climate Deaths"—urges criminal prosecution for homicide because the oil industry not only lied to the public about its carbon emissions but is legally complicit, harboring a culpable mental state, fully aware they have already killed thousands of people in the United States.

When these legal proceedings result in liability or guilt or both—and they will—the dollar costs will be enormous. The journal *One Earth* report of May 2023 begins with the conclusion that "human-caused climate change has long been acknowledged as essentially an ethical issue that threatens humanity and ravages the planet," then presents a methodological approach for morally based reparations and maintains ExxonMobil, BP, Shell, Saudi Arabia's state oil company Total, Chevron, and others among the world's twenty-one largest fossil fuel companies, which jointly owe more than $200 billion in annual climate reparations to compensate communities most damaged by their polluting business and decades of lies and greenwashing.[32]

There will be many more judicial pronouncements addressing the human right to a livable environment. By the time you read this, additional Our Children's Trust or similar lawsuits, environmental health and security reports, opinions, and other claims will have wound their way through the administrative, judicial, and political processes throughout the world. A friend of ours likes to think of the future as a child. By "future," he means the air we breathe, the water we drink, and the land where we produce our food—a vulnerable child requiring care. The Our Children's Trust litigation and other proceedings are introducing our friend's personal life thesis into public forums and discourse. He and many of those supporting these efforts believe we must care for the future as if it were a child. It's this caution and safekeeping that will allow humankind to provide a safe place for our children to carry on the American experiment of a democratic republic as well as the evolutionary experiment of human life.

CENTER FOR INTERNATIONAL ENVIRONMENTAL LAW

The mission of the Center for International Environmental Law (CIEL) is to use "the power of law to protect the environment, promote human rights, and ensure a just and sustainable society. CIEL seeks a world where the law reflects the interconnection between humans and the environment, respects the limits of the planet, protects the dignity and equality of each person, and encourages all of Earth's inhabitants to live in balance with each other."[33]

CIEL is a legal research and advocacy, education, and training organization with offices in Washington, D.C., and Geneva, Switzerland, that uses the power of the law to protect the environment, promote human rights, and ensure a just and sustainable society. The organization endeavors to rebalance the steep tilt in favor of corporate power by way of accelerating the transition away from fossil fuels, protecting forests and intact ecosystems, reducing toxic risks, and making trade safer. Its development is driven by human rights rather than by growth and extractive industry exploitation.

Hundreds of other organizations are also battling the Trifecta Crisis and those fostering it. While taking to the streets and personal choice are the most important things each of us can do, groups that pursue policy, litigation, and organized resistance to the status quo are critical too. Each of us can identify at least one group to support financially. CIEL suits Lucinda's and my interests, but many others are equally poised to focus the efforts of the willing.

Retirement to the Pitchfork Ranch has provided Lucinda and me with a unique opportunity to be part of the solution to the Trifecta Crisis. The most important lesson we've learned is that it's clear that everyone, everywhere, on any size property can become part of a natural climate solution in a meaningful way. Together, we can all not only lessen consumption and reduce greenhouse gas emissions, but we can also adopt a responsible way of life and pursue restoration that will help draw down the legacy load of atmospheric carbon.

The restoration work we have done on the Pitchfork Ranch is an important undertaking accomplished by just one couple in just one spot in just a short time, but it could be magnified many millions of times as part of a massive social movement to enable the survival of our and future generations as a collaborative, cooperative process. A campaign like this will be transformative. Both this moment and this scale are critical. It's time. We all have skin in the game.

CHAPTER NINE

Ecological Civilization

ALTHOUGH AMERICANS HAVE LONG professed an interest in improving land, restoration as an essential focus of environmentalism has been with us only since the mid-1980s. Habitat "protection has given way to restoration; the isolation of parks and wilderness has given way to integrated approaches to whole landscapes."[1]

Restoration ecology is now a discipline. As with most developments in human history, our views about conservation have gone through a lengthy process, an evolution of sorts, on the right way to live responsibly on the land. Habitat restoration is now firmly a part of our land ethic, yet, in view of the climate crisis, a far greater on-the-ground restoration effort is called for. To expand on language in Jonathan Rauch's *The Constitution of Knowledge: A Defense of Truth*—"given humans' innate tribal wiring, given our natural facility for hypocrisy and self-serving belief; given our many cognitive biases and our need to conform"—given current conflicting certitudes, divisiveness, and rancor, can we devise a land ethic, policy, or ideology that will channel self-interest, ambition, bias, and our other foibles to preserve civilization?[2] If the Trifecta Crisis and companion crises are to be overcome, a significant many of us must actively repair habitat; that work must be embedded in our forthcoming net-zero economy; we must develop a fossil-fuel-free way of life.

When *Homo sapiens* finally settled down after eons as hunters and foragers, fishers and fowlers, we became farmers and herders, an agricultural civilization. Human populations soared; God was discovered; and written history began. Several thousand years later we discovered coal, triggering the Industrial Revolution, followed by a two-century-long era in which we burned fossil fuels so aggressively that the planet has become critically overheated. We're now bogged down in what scientists call an existential climate crisis, not merely a threat to our way of life and place in the world, but literally a threat to human existence. Confronted with the current climate catastrophe, British environmental lawyer James Thornton and others like him envision a major civilization overhaul, what they describe as an "ecological civilization."

E. O. Wilson wrote, "There can be no purpose more inspiring than to begin the age of restoration, reweaving the wondrous diversity of life that still surrounds us."[3] In an ecological civilization, habitat restoration will attain its rightful place in our troubled world, at the pinnacle of importance. This transition entails far more effort in a considerably shorter period of time than the earlier passage from either hunter-gathering to farming and herding or from farming and herding to manufacturing. We have so little time; we must be in full transition by no later than 2030 to fend off the juggernaut threatening us.

Because environmentalism is the root of an ecological civilization, its evolution warrants review. Environmentalism is the result of a sluggish developmental process, "a long, slow revolution in values of which contemporary environmentalism is a consequence and continuation."[4] Restorationist and author William R. Jordan III writes that progress in ecological thinking has had three phases, reduced to their simplest form: landscape as *colony*, habitat as *sacred space*, and, last, landscape as *community*. Climate breakdown and our wholesale inertia in response to the crisis requires we think differently about ecology and environmental ethics. This third phase of landscape as community would best serve life if it were to evolve into thinking of habitat as *survival*. The idea of survival is a way to thwart the inaction and benign neglect, to instigate a willingness to forgo immediate gratification and convenience, and to take a multigenerational look at what is meant by progress.

In keeping with Jordan's three phases of ecological thinking—the first was colony—concern for the environment initially entered our thinking in a serious way with the writings of Ralph Waldo Emerson and Henry David Thoreau. They believed that human beings have a special relationship with nature. Emerson and Thoreau shared a love for the natural world, a belief that all living things have rights that humans should recognize and that we have a responsibility to respect and care for nature rather than merely exploit it. Thoreau reasoned that "every creature is better alive than dead . . . and he who understands it aright will rather preserve its life than destroy it."[5] Emerson, Thoreau's mentor, argued "that humans should seek to develop a special relationship with the universe in order to regain a sense of 'self,'" which he felt had been lost in the modern world.[6] Thoreau was a surveyor, engaged in commerce. And Emerson also endorsed commerce, writing that capitalism was about "doing well as a result of doing good."[7] Despite their unambiguous affection for nature, their thinking about the natural world included the perspective that it was also a resource in service of humankind.

In this early phase of conservation, the value of a given place was grounded in utilitarian thinking, as in its worth as a source of goods and services. The conservation movement saw itself as master and focused on habitat in terms of providing food, fuel, and other consumables such as clean water and lumber. Conservation had a biblical bent, derived from Genesis: "Be fruitful, and multiply, and replenish the earth, and subdue it: and have dominion over the fish of the sea, and over the birds of the air, and over every living thing that moveth on the earth."[8]

Landscape was thought of as a colony that existed for the benefit of man. The frame of reference for this thinking was industry: the fur trade, logging, open-range cattle ranching, mining, and irrigation farming. The philosophy of the first chief of the U.S. Forest Service, Gifford Pinchot, was one of "wise use," the idea of sustainable harvests of natural resources from public lands. He coined the term *conservation ethic*, which, for him, meant "the art of producing from the forest whatever it could yield for the service of man."[9]

John Muir initiated an evolutionary second phase of conservation thinking: wilderness as sacred space to be esteemed and preserved. He had little interest in commerce; rather, Muir's ideas were based on a spiritual perspective of preserving land. According to Howard Zahniser, an activist with the Wilderness Society, in language incorporated into the 1964 Wilderness Act, wilderness was "an area where the earth and its community of life are untrammeled by man, where man himself is a visitor who does not remain."[10] Wallace Stegner wrote in his influential 1960 "Wilderness Letter" urging preservation of the great outdoors: "we need wilderness preserved . . . even if we never once in ten years set foot in it. . . . It is important to us when we are old simply because it is there—important, that is, simply as an idea." He foresaw sorrow and grief if wilderness was lost: "Something will have gone out of us as a people if we ever let the remaining wilderness be destroyed. . . . We simply need that wild country available to us, even if we never do more than drive to the edge and look in."[11]

Although more a writer of the third phase of conservation thought, Terry Tempest Williams writes that wilderness is necessary to be authentically human, "To be whole. To be complete. Wildness reminds us what it means to be human."[12] Late in life, A. B. Guthrie Jr. expressed a similar view of the specialness, even sacredness, of wilderness: "I am too old to go there again, but I rejoice that it exists for others. It is its own excuse for being the wilderness. Its wonders should be passed on pristine."[13] The emphasis during this second period was on "preservation" of a sacred place from which humankind was separate and apart.

Environmentalism next moved into the revered Aldo Leopold phase, which understands conservation in terms of a community to which we all belong, a world of sentient beings where all species make up an interconnected and interdependent community. Leopold was exhaustive in his inclusiveness. He famously wrote, "The land ethic simply enlarges the boundaries of the community to include soils, water, plants, and animals or collectively: the land."[14] He initiated changes in fundamental assumptions about natural resources and the nature and purpose of ecological studies, changes without which the discipline of conservation biology could not have emerged. His view of the moral consideration of land as community is the starting point for almost all discussions of environmental ethics.

Leopold's classic essay "The Land Ethic" in *A Sand County Almanac* is the most widely cited source in the literature of environmental philosophy. Leopold's "land ethic" formulated a morally linked relationship with Earth. American philosopher J. Baird Callicott writes, "The 'Land Ethic' in *A Sand County Almanac*, though to one degree or another anticipated by Thoreau, Darwin, and Muir, is the first self-conscious, sustained, and systematic attempt in modern Western literature to develop an ethical theory which would include the whole terrestrial nature and terrestrial nature as a whole within the purview of morals."[15]

Essential to this current phase of conservation thinking is the belief that we have an ethical obligation to treat the rest of life as we treat ourselves: "Leopold developed a philosophy that accorded moral value not only to other species, but also to inanimate features of the landscape such as mountains and rivers, to natural processes such as cycling of water and nutrients through ecosystems and, on a longer time scale, to the formation of new species through natural selection."[16] For Leopold, the conservation movement existed because the ongoing "interactions between man and land were too important to be left to chance."[17] He wrote, "It is inconceivable to me that an ethical relation to land can exist without love, respect and admiration for land, and a high regard for its value. By value, I of course mean something far broader than mere economic value; I mean value in a philosophical sense."[18]

He maintained responsible land-use conservation sprang from a sense of individual responsibility for the general health of the land. He saw three options to remediate nature's exploitation and ruin, three means of control: legislation, self-interest, and ethics. Leopold harbored the idea—while acknowledging it as a pipe dream—of "a new fellowship to the land, a new solidarity in all men privileged to plow, a realization of Whitman's dream to '*plant companionship as thick as trees along all the rivers of America*.' What bitter parody of such companionship, and trees, and rivers, is offered to this our generation!"[19]

Leopold's ideas were "grounded in an appeal to four values: the integrity and stability of the land community, its beauty, and—centrally—community, and the land conceived as a community of organisms."[20] When he and his wife, Estella, leased and then purchased an eighty-acre piece of worn-out land along the Wisconsin River in Sauk County, Wisconsin, his family, students, and friends did some of the first conservation-inspired habitat restoration. As an early practitioner, Leopold pursued the restorative ideal that was deeply embedded in his thinking about sick land. Repair or restoration of habitat is a core part of today's conservation ethic, allowing us—not only the farmer or rancher—to experience an engaged, participatory relationship with the natural world.

Many readers of this book will be hampered, as I am, by an assumption in Western thought, a core element of our cultural inheritance. A spirit-nature dualism is ensconced in Western cultures, steeped as they are in the Abrahamic traditions. At its most basic, our thinking goes like this: there is a true and universal human nature, and it recognizes humans as superior to and apart from the natural world—superior because humans were created by God, in his image. God made humans both different and superior to other species in that humans have a soul that is subject to salvation.

One of the most all-encompassing assessments—and there are many—championing humankind's specialness and superiority over all manner of life and its dominion over Earth is Bishop Samuel Wilberforce's challenge to Darwin's evolutionary theory of natural selection. It was presented at the 1860 annual meeting of the British Association for the Advancement of Science, about seven months after the publication of Darwin's *The Origin of Species by Means of Natural Selection.* Wilberforce declared, "Man's derived supremacy over the earth; man's power of articulate speech; man's gift of reason; man's freewill and responsibility; man's fall and man's redemption; the incarnation of the Eternal Son; the indwelling, of the Eternal Spirit,—all are equally and utterly irreconcilable with the degrading notion of the brute origin of him who was created in the image of God."[21]

The book *Religion as We Know It: An Origin Story* by Jack Miles, University of California–Irvine distinguished professor emeritus of English and religious studies, contends that followers of Jesus early on acquired the habit of thinking of their religion as a separate domain from their day-to-day lives, joining the flock that required converts to abandon their worship of Roman gods. Miles maintains this same story of "separateness" continues today in the hearts and minds of individual religious, as well as irreligious men and women: "Christians did indeed acquire very early and thereafter never entirely lost the habit of thinking of their religion as a separate domain. Once this is conceded, it should come as no great surprise that as a corollary of this habit, they should have adopted early and never entirely lost the habit of thinking of other religions, rightly or wrongly, as similarly separate domains."[22]

Recall the long-held centrality of Earth-centered cosmology to the Holy Mother Church in keeping with the belief in heaven and hell, the former an eternal and incorruptible paradise. The "sins" and suffering of early scientists were widespread: Giordano Bruno and Galileo were burned at the stake and confined under house arrest for life, respectively. Growth, decline, and decay belong to Earth, sanction for the sins of our forebears. Nature is where we sin. Believe and repent and salvation will be yours in heaven with God forever—yes, immortality.

The fundamental flaw in the tenet of Christian salvation—and its implications for our beleaguered planet—is it means humans do not belong in this world; rather, our final resting place is to go home and be with God. According to this view, "Humanity's real home does not lie among the rest of creation but rather with God in heaven; [our] true home and destiny lie in the kingdom of God," writes scholar Anna L. Peterson.[23]

Academic and secular humanist Robert Wright calls this the "morally contingent afterlife" in which "your actions in this life have a direct effect on your experience in the next." In

Wright's words, "it makes a blissful afterlife contingent on your moral fiber."[24] The sine qua non is compliance with divine rule. Peterson addresses this fundamental separation in her *Being Human: Ethics, Environment, and Our Place in the World*: "By emphasizing humanity's visitor . . . status in the created world, salvation reinforces the division between humans and nature decried by many critics of Christian anthropocentrism."[25]

In keeping with Wright's, Peterson's, and Miles's thinking, a good number of scholars agree that the arrival of the earliest monotheistic traditions initiated "a shift in the source of sacredness from the world up to the heavens."[26] Half a century ago, Wendell Berry had similar thoughts, writing that this perspective "has encouraged people to believe that this world is of no importance, and that their only obligation in it is to submit to certain churchly formulas in order to get to heaven." In "pursuit of 'salvation,' . . . the Heaven-bent have abused the earth thoughtlessly, by inattention, and their negligence has permitted and encouraged others to abuse it deliberately."[27]

The secular Western mind also views humans as separate, unique, and superior because of our consciousness of being human or because we possess a rational mind, language, culture, or some other intellectual "something" that other animals don't enjoy. Plato and Aristotle saw reason as the wellspring of virtue; we are moral beings because we are rational beings. Samuel Wilberforce's view of man as supreme was grounded in both reason and redemption. Whether it's the "eternal soul" (religious) or "rational mind" (secular) or, as with the bishop, some combination of the two that sets us apart from nature, this dichotomy between humans and nature helps explain our misuse of the planet's resources.

This also helps explain why Leopold's environmental ethic and practices have failed to gain sufficient traction in the American mind to contend with the Trifecta Crisis. His land ethic centered us, habitat, and its other occupants in an Earth-based community where humans understand their time here is temporary, hoping to eventually make their way to an eternal life in heaven. This reasoning parallels the elemental flaw at the core of Western thought: "Leopold's conception of community . . . rests ultimately on an appeal to altruism. This may seem a forgivable weakness, especially since it is one that Leopold shares . . . with such luminaries as Gandhi, Martin Luther King and the Buddha. Yet clearly there is something missing from Leopold's land ethic."[28]

The shortcoming of altruism unmasks the deficiency in Leopold's ethic. This flaw is compounded by Leopold's recognition that a land ethic could not take hold without love, respect, and admiration for land. Piling the requirement of love and altruism atop our ending life elsewhere—as in heaven—his land ethic never had a chance, and it is certainly inadequate to undergird humankind's challenge to contend with the Trifecta Crisis. Despite humankind's distinction as "the extreme biological outlier . . . among mammals . . . [for] our astonishing degree of altruism: our kindness towards other members of our species," reliance on the practice of selfless concern for the well-being of others and love has left us with a hapless environmental ethic.[29] The current state of affairs confirms that humankind, in particular, those engaged in a consolidated, corporate, capitalist economy, has never loved the land. Leopold's land ethic has fallen short of its fundamental objective: the preservation of life on the planet. The flaw in his reliance on altruism and love ignores the inherent, pervasive conflict between self and community.

When survival becomes the priority, it compels the fusion of a more risk-averse self with the whole community of life. In these crises, the conservation ethic will no longer be held back by thinking of landscape as *colony*, as *sacred space*, or as *community*, grounded in altruism. Leopold's ethos becomes compelling only when landscape is thought of in terms of *survival*. Because it is my principal aim, allow me to reiterate my hope that the idea of survival will thwart the inertia and indifference to these crises, instigate a willingness to forgo immediate gratification, take a multigenerational view at what is meant by progress, and define progress as more than growth. Emphasizing survival leads us to reimagining and reconfiguring how we live. It creates the need for a fresh, thoroughly different mental language that will help us change the relationship between humans and the world around us. Looking at our future through the lens of survival can give us a new mind set, a crises-centered outlook needed to reverse the tragedy of the commons, the crime of our unrestrained consumption of Earth's nearly depleted natural resources.

Admittedly, the conflict of interest between self and community in altruism is at play with the substitution of survival for community in Leopold's land ethic. Yet the inherent human tendency to engage in disinterested and selfless concern for the well-being of others pales in comparison to the survival instinct, the desire to stay alive and keep our children safe. Survival is a far more compelling sentiment than any natural tendency toward doing the right thing for others. Presbyterian minister, American journalist, author, and freelance war correspondent Chris Hedges: "For those of us who have spent years in wars, it is the suffering of children that most haunts us. If, as a society, we see that our principal task is the care of children, of the next generation, then the madness of the moment can be dispelled." According to historian and conservationist Curt Meine, "Something really difficult is on the way. . . . If we are going to survive and to thrive in whatever landscape the world offers us in the decades ahead, we must learn to speak respectfully to each other, to listen to each other, to take into consideration the fate of each other's children."[30]

Darwin's and E. O. Wilson's thinking about the virtuous joining forces with others of virtue in groups, outcompeting the less virtuous, is emboldened when their alliance is grounded in survival. Survival of oneself and one's children as members of a threatened group compels collaboration and the recognition that preservation of the planet and all its life-forms are what's at stake. Lest we forget, it's always amoral "to survive at the expense of somebody else," and in this case, the group is global.[31]

This notion of survival, while given too little emphasis, has been with us a long while, particularly in the pantheon of love letters to Earth. In 1985 Wallace Stegner wrote that Leopold's *A Sand County Almanac* was "almost a holy book in conservation circles . . . one of the great love letters to the natural world."[32] Stegner's own 1960 "Wilderness Letter," used to introduce the bill establishing the National Wilderness Preservation System in 1964, rests in that sanctuary too. Love letters to Earth persist, and they're always grounded in survival. Vietnamese peace activist and Buddhist monk Thích Nhất Hạnh made clear the survival risks of what we're doing to our home in his 2013 book, *Love Letter to the Earth*: "there is no difference between healing ourselves and healing the planet," and "caring for the environment is not an obligation, but a matter of personal and collective happiness and survival." He maintains that we "have to change our whole relationship with the Earth," "walk differently and care for her differently," and develop a loving relationship "with the Earth if the Earth is to survive, and if we are to survive as well."[33] Leopold, Stegner, and Nhất Hạnh each recognizes what has not yet caught on: our survival is at stake if we don't quickly and dramatically change how we live on Earth.

Amitav Ghosh argues, "In an ironic twist, the individual conscience is now increasingly seen as the battleground of choice for the conflict is self-evidently a problem of global commons requiring collective action; it is as if every other resource of democratic governance had been exhausted leaving only this residue—the moral."[34] Yet faith or hope in the politics of moral sincerity also falls into the altruism trap. There is a tension between moral altruism and amoral selfishness, absent when pursuing the survival of our offspring. The 2018 IPCC report's call for "rapid, far-reaching and unprecedented changes in all aspects of society" was an appeal for radical, structural change grounded in our continued survival, staying alive.[35] Yes, if human civilization is to avoid collapse and maintain food security and ecosystem integrity, we must recognize that our survival depends on collective action by the entirety of Ghosh's global commons.

Hidden in the immediacy of the climate crisis lies the potential to resolve the conflict built into relying on altruism. The loss of biodiversity and soil and the climate crisis are creating shortages of food and water and will increase the risk from cross-species contagions that leak into civilization, as we've seen with the COVID-19 pandemic, in which nearly 7 million humans have perished—with some estimates as high as more than 18 million.[36] A first-of-its-kind study in *Nature Climate Change* found the emissions of greenhouse gases have been exacerbating 58 percent—218 of the total 375—of pathogenic diseases known to have impacted humans.[37] The authors maintain the numbers make it clear that with so many diseases and so many pathways, there is no way humankind can fully adapt. This onslaught of heat and disruption overwhelming Earth compels us to see how Leopold's thinking has not been sufficiently persuasive to influence enough people to recognize and avoid the consequences of the Trifecta Crisis. At the cliff's edge, substituting *survival* for Leopold's *community* eliminates the conflict deep-rooted in altruism. The steep descent ahead just may awaken us.

It was a surprise to find the notion of survival foreshadowed my thinking by half a century in Wallace Stegner's obscure 1969 essay titled "Conservation Equals Survival." Although he was thinking in terms of the atmosphere's thinning rather than warming or "ecological disruption, depletion, pollution, [and] the shrinking of healthy open space," he nonetheless was onto the fact that "fossil energy is the worst discovery man ever made, and his disruption of the carbon-oxygen cycle is the greatest of his triumphs over nature." Stegner cites Lamont Cole of Cornell, who understood that "our large scale burning of fossil fuels [endangers] the atmosphere in other ways than pollution."[38] Stegner suggested the solution to what ails us. Although far too quaint for current threats, it "could begin with the little individuals, the kind of people many would call cranks, who insist on organic grown vegetables, and unsprayed fruits, who do not pick the wildflowers, who fight against needless dams and roads. For that is the sort of small personal action that a land ethic suggests. Widespread enough, it can keep men from moving mountains."[39] As with Leopold, although the climate crisis was neither's worry, Stegner knew "we [could] destroy ourselves . . . through continuing abuse of our environment . . . [and, as Stegner made clear,] our disruption of the carbon-oxygen cycle" was an atrocity.[40] Both Stegner's and Leopold's recognition that humankind was capable of destroying the livable world and extinguishing itself strikes me as the most significant, most prescient insight either of them ever had. And they both, as Stegner noted in his "Survival" essay, rightly saw that the solution to these wrongs was "personal action," an action agenda that "can keep men from moving mountains" and destroying the planet, the Voice of the Streets.

The deficit in humankind's thinking, our inability to come to grips with challenges and threats, has long been known. In 1758 French philosopher, writer, and composer Jean-Jacques Rousseau—said to have conquered "the whole of human history" with his establishment of the "hunter/farmer/industrialist" story of social evolution, as well as inspiring the French and American Revolutions—wrote in his "Letter to M. d'Alembert on Spectacles" of a philosophical idea that resonates yet today: "People think they come together in the spectacle, and it is here that they are isolated."[41] His objective was for us to understand "spectacle," as in a theatrical performance, suggesting when we attend the playhouse as mere spectators, we see things at a distance—entertained and disengaged; we observe suffering on stage in comfort while detached. This remoteness instills inaction and clouds our world with unresolved flaws, leaving us isolated in our togetherness. We do talk to others about what we've seen, but these are merely verbalizations, not actions that can lead to activity and engagement. As a consequence, we lose our compassion and our capacity to actively engage in addressing concerns, helping others and taking to the streets.[42]

Although Rousseau was using the venue of theater to make his case that our participatory attributes have atrophied, his reasoning applies far beyond the playhouse. In keeping with his thinking, rather than using theater as spectacle, think in terms of society itself as spectacle. In 1967 another French writer and political theorist, Guy Debord, published *The Society of the Spectacle*, in which he argued that the history of modern social life can be understood as an economic domination, downgrading existence from *being* to *having* and from *having* to merely *appearing to have*.[43]

Rather than being authentically human, our current way of life has reduced us to, at best, merely having the characteristics of being human or, at worst, merely appearing to have the characteristics of being human. A number of writers have recognized the similarity between Rousseau and Debord's thinking about the spectacle as a vehicle of alienation. Even de Tocqueville wrote about "the possible degeneration of the modern individualism diagnosed by Rousseau."[44] Constitutional standards and political norms have broken down, the prevailing way of life has diminished the human spirit, our lives in the spectacle have led to a reluctance to engage the world in a meaningful, consequential way. With the Trifecta Crisis bearing down, best we floor the throttle of change, abandon the yammering about events and people and focus on well-known solutions to these crises.

How do we abandon the conveniences of fossil fuel–rooted modern life, the spectacle, and reach for something more? How do we adopt different goals, develop a new language, a responsible way of life and pursue survival? We addressed strategies that will save us in chapter 8 and we'll look at how to ramp up our efforts, abandon the spectacle, and reach for something more in chapter 12. But the short answer is that we must develop an integrated, scientifically based, and enlarged sense of self that expands beyond our own skins and encompasses the survival of children and the rest of life on the planet. There's a belief of Indigenous peoples that serves one of the core ideas of this book, the survival of our children: "Your life doesn't belong to you. You're only here to ensure the coming generations."[45]

This can be thought of as redefining self-interest as part of an ecological civilization. The idea of the "gradual force of persuasion," more or less a civil standard, is obsolete in view of the crises' immediacy. Compromise is said to be central to democracy, but we don't have the time for caution, for the middle ground. The place to begin, to return to a philosophy that matters, requires the abandonment of meaningless individual and media mumbling. Mistakenly attributed to social activist and former First Lady Eleanor Roosevelt, the maxim "great minds discuss ideas; average minds discuss events; small minds discuss people" is a pejorative opinion derivative of class.[46] Yet recast in humbler verse, it reaches the core of many of today's shortcomings: "Responsible minds discuss ideas; distracted minds discuss events; and immature minds talk about people." A fitting example of immature minds mired in the swamp of celebrity is Kylie Jenner having 387 million Instagram followers. This era of low-effort thought must end, and we need to revitalize our shallowed sense of what is possible. The core idea is to begin conducting ourselves as "residents and shapers, but not masters and makers," of the natural world and its vast universal web of life's profound interdependence, a new ideal of survival citizenship.[47] We are not *separate from* but *part of* something in which our survival is at stake. The collaborative, scientifically based street activist adopts a personal strategy focused on what can be done now to assure the survival of future generations.

If Aldo Leopold were alive today, in the face of the immediacy of these crises, it's likely he would rethink his reliance on community and recognize survival as the best way to accelerate the transition out of our industrial civilization and into an ecological civilization. There is certainly nothing untoward of a land ethic dependent on individual responsibility grounded in altruism; it's simply not sufficiently compelling in the face of these crises. Taking the high road, Leopold maintained conservation was grounded in a deep sense of individual responsibility for the general health of the land. And while altruism speaks to the best in us, it fails to take into account the false dichotomy between the self and other life-forms and nature generally. By use of the phrase "false dichotomy," my intent is to emphasize that humans and other life-forms are far more alike than different, that understanding humankind as separate from the rest of life fumbles the ball. Acknowledging this flaw, a land ethic understood as essential for the survival of our children is far more compelling.

As with Wallace Stegner, although overheating the planet may have been no concern in Leopold's day, there are textual indications he was thinking broadly enough to include it. In 1933 Leopold delivered the fourth annual John Wesley Powell Lecture to the Southwestern Division of the American Association for the Advancement of Science in Las Cruces, New Mexico, where he said,

> In short, the reaction of land to occupancy determines the *nature and duration of civilization*. In arid climates the land may be destroyed. In *all climates* the plant succession determines what economic activities can be supported. Their nature and intensity in turn determines not only the domestic but also the wild plant and animal life, the scenery, and *the whole face of nature*. We inherit the Earth, but within the limits of the soil and the plant succession.[48]

Leopold was thinking about something larger than the "land sickness" that underscored so much of his philosophy. His ethic applied to "all climates," plural, to "the whole face of nature," and to the "nature and duration of civilization." Who thought about the end of civilization in his day? While Leopold presumably had no awareness of the risk posed by an overheated climate, he too could see that "human beings collectively have acquired the power to destroy the integrity, diversity, and stability of the environing and supporting economy of nature."[49]

Leopold counsels: "the 'key-log' which must be moved to release the evolutionary process for an ethic is simply this: quit thinking about decent land-use as solely an economic problem. Examine each question in terms of what is ethically and esthetically right, as well as what is economically expedient. A thing is right when it tends to preserve the integrity, stability, and beauty of the biotic community [and] wrong when it tends otherwise."[50]

Setting the stage in an early essay for the eventual adoption of a land ethic, Leopold subordinated individual members of the biotic community for the greater good of the whole. According to J. Baird Callicott, "Leopold speculatively flirted with the intensively holistic superorganism model of the environment as a paradigm pregnant with moral implications."[51] Writing in terms of "the indivisibility of the earth—its soil, mountains, rivers, forests, *climate*, plants, and animals, and *respect[ing] it collectively* not only as a useful servant but as a living being," or "the earth's parts—soil, mountains, rivers, *atmosphere*, etc.—as organs, or parts of organs, or *a coordinated whole*, each part with a definite function," he prioritized the whole Earth as "an organism possessing a certain kind and degree of life, which we intuitively respect as such," over the interest of any individual or group.[52]

As Callicott recognized, Leopold's "land ethic not only provides moral considerability for the biotic community per se, but ethical consideration of its individual members is preempted by concern for the preservation of the integrity, stability, and beauty of the biotic community. The land ethic, thus, not only has a holistic aspect; it is holistic with a vengeance."[53] The term "holistic" reminds us of Leopold's continual use of the word "whole," as in medically treating the whole person and all aspects of our being. Using Leopold's terminology, the "reaction" of "the whole face of nature" to our "occupancy" describes the current crisis. It's been written by more than a few scientists that humans have broken the Earth, that we are threatening the nature and duration of civilization.

Thinking in terms of survival turns ideas of uncaring self-interest on their head: in order to coexist in my community beyond the short term, I cannot help myself and my children without helping you and your children and the whole of nature. And the same goes for you and everyone else. For too long, we have understood nature as separate and have reduced it to a storehouse—"Paul Sear's warning that nature is not an 'inert stockroom'"—a mere means in service of ourselves, a perception that has impoverished us and put civilization at risk.[54]

In the context of climate chaos, the rapid extinction of species, and the deterioration of Earth's biosystems, survival is possible only if we think of this enlarged sense of self as an "ecological self," a member of an ecological civilization. This survivalist, ecological self summons us to ground our choices in constraints other than duty to God or altruism. Ethical citizenship incorporates the common sense carrying on of our families, of our species, of life generally. The task at hand is learning to think horizontally, as in helping Earth, rather than vertically, as in seeking salvation and entry into heaven. Think in terms of action, as beavers do, another animal besides us with the capacity to significantly reshape the environment and serve civilization as a vocational restoration practitioner. Think watershed: work to help these critical spaces, hold on to their soil and rainwater, and reverse the misuse allowing so much soil to wash away. Think plants: plant seedlings and work the land to enhance flora and its natural, life-giving capacity to provide us with oxygen and capture carbon. If we are to mature into an ecological civilization, we need to begin imagining a different future and to think in terms of survival based on an environmental ethic that insists we protect our home. In order to have a tomorrow, we must join forces with other people and other species. Survival is a collaborative, cooperative process. This will be a tectonic shift, arguably initiating a new way of thinking about human life, all life, an outlook that highlights to an unprecedented degree the value of the interconnected natural world and the power of rational inquiry into science, the efficacy of the survival instinct, and the future of our children and our children's children.

CHAPTER TEN

The Empire Zinc Mine Strike and *Salt of the Earth*

A THIN SLICE OF PITCHFORK Ranch history involves the triggering of the most significant social and political event in Grant County, New Mexico, history. This retelling is not a chance, place-based inclusion in this book: the earlier owner of the Pitchfork Ranch triggered these events. It's an essential lesson in confronting an overwhelmingly formidable status quo and an encouragement on how to tackle the challenges posed by the Trifecta Crisis in the short time we have remaining. It underscores the importance of the Voice of the Streets. Civil rights leader and major figure in Mexican American politics during the 1950s and 1960s Vicente T. Ximenes maintains the strike "offers a model for direct action, organization, collective activism and deliberative processes."[1]

The year was 1949. Back from the war, recently arrived in Grant County, New Mexico, Clinton Jencks headed Local 890 of the International Union of Mine, Mill, and Smelter Workers (commonly referred to as Mine-Mill), whose members were employed by the Empire Zinc Company mine in the town of Hanover near Silver City. Bartley McDonald, co-owner of the Pitchfork Ranch, was the Grant County sheriff. What occurred in Grant County has happened often elsewhere but only once here. For fifteen months from October 1950 until January 1952, miners and their wives challenged four of the five most powerful players in the county: the Empire Zinc Company, Grant County government, the legal system, and law enforcement authority.[2] Cattle ranching was the fifth player, and with cattleman and Sheriff McDonald in the mix, the conflict was between the strikers and nearly everyone.

When Clinton and Virginia Jencks arrived in Grant County, it was well known communists had infiltrated the union. Coupled with racism, "red" antagonism played a central role in the troubles that would soon engulf Grant County. Discrimination in pay, job assignments, housing, sanitation, a no-strike contract clause, and other disparities in the treatment of Hispano and Anglo miners were the backdrop for the events of April 30, 1949.

It was one of the first warm days of the year, a typical Saturday night at the Fierro Nite Club in Bayard, Grant County: Mexican patrons crowded in, dancing and drinking. The details of the fracas are not necessary for our purposes, but, in its simplest terms, a brawl began when two Grant County sheriff officers arrested union member Rubén Arzola for drunk and disorderly conduct as he and a friend walked out of the dance hall. Arzola's friend hastened to defend him, shouting in protest, causing other patrons to pour outside. He too was arrested. A large crowd gathered; hostility, turmoil, and confusion followed; shots were fired by one of the officers; people scattered; several patrons and Arzola were wounded; rocks were thrown at the officers; and their car was overturned in the melee. The arresting officer was notorious throughout Grant County for his hostility toward Mexicans. It was bedlam. Accounts differed, and resentment boiled. More arrests of union members followed.

The next day, Local 890 head Clinton Jencks went to the sheriff's office but failed to get answers. On Monday, Jencks returned, but Sheriff McDonald told him only a minister, lawyer, or doctor could meet with anyone held in his jail; he rejected Jencks's plea that he was "counsel" for the prisoners. McDonald stated, "I told him he wasn't a lawyer and, as far as I was concerned, [he] could quit meddling in the case." The sheriff's department continued to arrest miners, the total eventually reaching ten men. On Tuesday, Jencks returned, this time with attorney C. C. Royall Jr., to again ask about the cases and why they were not yet on the docket of the justice of the peace. Jencks, members of the union, and their families knew detainees were not being timely processed. McDonald claimed that Jencks called him a liar. Jencks denied it, claiming he just wanted to see that the prisoners got an even break. As Jencks put it, "McDonald got mad and hit me."[3] McDonald dropped Jencks in a one-punch fight, igniting the Hispanic community.

They finally had enough. Many meetings followed the McDonald-Jencks incident, one as large as three hundred people.

Community activism spiked. Outsiders stepped in. The newly established American GI Forum chapter eventually supported the miners and their families by providing food, clothing, and supplies. Other reform efforts reached into a variety of areas: Felix Salinas, the forum's legal adviser, appeared before the Silver City School Board and was successful in getting the gerrymandered school districts redrawn. Forum leaders organized Mexican participation in local government, cultivating civic literacy, and public education.

My sense of connectedness to these events deepened when I realized the startling link to my former working world. The man Bartley McDonald decked in the Grant County sheriff's office was the defendant in the famed U.S. Supreme Court case establishing the Sixth Amendment confrontation rights of an accused in a criminal prosecution. Every law school student knows the case, *Jencks v. United States*, establishing a criminal defendant's right to inspect prosecution witnesses' prior statements and to cross-examine the witnesses before the trier of fact, usually a jury.[4] Taken for granted today, the decision is among the most important Supreme Court cases in American criminal jurisprudence, second only to *Miranda*, the right to council case.

Although Bartley McDonald was a highly visible fish in the small Silver City pond and his antics were the trigger that led to the strike, the important Supreme Court case, and film that followed, he turned out to be little more than a bit player in a historically important series of events in which the courageous participants were Clinton and Virginia Jencks and the miners' wives. The lessons learned and the precedent established reflect the character and ideals of the Jenckses, the strength of the miners' wives, the importance of "action," and the power of a mobilized people, the Voice of the Streets.

To contend with the structurally entrenched racist policies of the Empire Zinc Company and the community at large, Clinton Jencks brought an unfamiliar point of view to the region of well-established norms. He introduced an action-based means of promoting change. In terms of harshness, the political split in America today—inflamed as it is by ideological anger, moralistic certitude, and the absence of a shared narrative, which makes compromise tantamount to treason—is similar to the world Jencks and his wife walked into when they began employment with Local 890.

The Empire Zinc Company mine strike, surrounding disputes, and even the larger context of the time—the Cold War, the Red Scare, McCarthyism, unions, equal pay, workers' rights and dignity, Mexican American civil rights, and, most surprisingly, gender equality and feminism—were memorialized in a film about the strike that has been shown worldwide, *Salt of the Earth*. It was the first and only American movie to be blacklisted in the United States. The tortured history of the film illustrates a number of important points: the fear that paralyzed America in the days of the House Un-American Activities Committee inquisition, McCarthyism, and the implications of a "temporary inversion of sexual division of labor . . . help[ed] dismantle the dual-wage system [and] open[ed] cinema to realistic portrayals of working-class life."[5]

Filmed in 1953 near the location of the famous strike and soon after the fifteen-month walkout ended, *Salt of the Earth* was banned in America and, for many years, rarely shown. The film was initially played in only thirteen theaters throughout the nation and was not widely available in this country until the mid-1960s. It came to be a civil rights classic, an epic class struggle, often cited as among the most important films of the twentieth century. It's been screened in fifty to sixty different languages all over the world. There are also books devoted to it and to the strike itself, and two biographies of Jencks.[6]

The strike and movie were revisited in the Silver City Museum's 2004 four-day colloquium Bringing *Salt of the Earth* Home: A 50th Anniversary Symposium. The conference was an effort to provide a comprehensive survey of the strike and movie while the participants in these events were still alive. It was held to commemorate the strike, the film, and the wide-ranging implications of both, while continuing to strive for an understanding of what has proven to be a series of controversial, divisive, and bitterly painful events. Locals—both Hispanos and Anglos—typically don't talk about the strike and movie. Although racial discrimination is an acknowledged blemish on Grant County's past, often hinted at but seldom owned up to, the topic of the strike has long been an edgy issue that, for just about everyone, remains under wraps. Even though the strike and film still hold sway in the American history of political protest, their significance is under appreciated in Grant County.[7] Long-simmering resentments, bitterness, and hard feelings bubble below the surface, and race-based feelings persist like scabbed-over wounds that are ignored but not forgotten, sore spots in the mind that won't bear touching.

The learned practice of loathing those with skin of a different hue was entrenched and very much in play in the mid-twentieth-century Southwest. The most obvious example of this bigotry was the miners' long-standing, much-maligned, and demeaning dual-wage system. Mexican American miners were paid significantly less than Anglos, along with being subjected to the practices of dual-pay windows and segregated lunchrooms, showers, changing quarters, and attendant facilities. These racist practices remained from a time when Anglos were paid in dollars while Hispanos were paid in scrip that could be used only at the company store. Turn-of-the-century pay books from the larger nearby Santa Rita Mining Company are still available and document the pay question at the heart of the strike: any miner with a Hispanic surname was paid two dollars a day; Anglo miners were paid three dollars.

This was a world not unlike that of the "separate but equal" drinking fountains and off-limits lunch counters of the Jim Crow south. It was a time when children had their hands smacked with a wooden ruler for speaking Spanish. Lucinda and I have a Hispanic friend whose eighty-year-old father was a Silver City Lincoln Elementary School student, who, along with other Mexican American kids, was stripped and hosed down before school, thought too dirty to be with White kids. Even today, our friend's father distrusts Anglos and will engage us only when necessary. He wouldn't meet with me. More than half of the miners' homes, provided by the Empire Zinc Company, had no running water. Outhouses were the norm. Empire representatives maintained that the topic of indoor plumbing was outside the scope of the union's bargaining rights.

Far more serious than the plumbing issue is the fact thirty miners were killed in mining accidents in Grant County between 1946 and 1950, and hundreds more were injured.[8] With more than 90 percent of the Grant County union labor force being Mexican American and doing the underground and hardest jobs, it is clear on whom the bulk of mine-related deaths and injuries fell.

The left-wing Mine-Mill Local 890 arrived in Grant County in the 1930s with an already established militant approach to labor demands and a transparent connection to the Communist Party. The prewar push for metals created the opportunity for Local 890 to establish a toehold in Grant County, as the labor shortage and other factors obscured anti-union and racist sentiments. Throughout the 1940s, workers and organizers made strides toward ending the dual-wage system—except for workers at the Empire Zinc Company mine, which had the dubious distinction of being the only mining company in New Mexico continuing the dual-wage system at the time of the strike. Only twelve of the ninety-two unionists who joined the Empire mine walkout were Anglos, and there were no Anglo participants on the picket line.[9]

With the threat of communism in full swing, mining companies, managers, and those who either spoke for or were indebted to mining interests exploited the national anti-communist sentiment to which Mine-Mill was vulnerable. The union's tolerance of leaders and members who were outspoken in their favorable leanings toward the Communist Party was well known. Yet the unions knew management was far less concerned about loyalty to country than exploiting Cold War sentiment as a means to reinstate the racist order and its associated cheap labor that dominated prewar America. What mattered to union members was equal pay for equal work and respect. The granddaughter of union photographer José Carrillo gave me a photo of the strike area with a sign: "We want equality with other workers, no more—no less." The rank and file had little interest in political ideology. Mine-Mill former vice president Orville Larson remarked that civil rights and workplace equality were in the union members' interest and "the communist issue didn't matter a hoot." Union radicals saw the Communist Party as an ally in the advancement of social justice and racial equality through socially engaged unionism.[10]

It's difficult to understand the strike, the violence surrounding it, and the film that told the story of the strike without an appreciation of its larger context: the Red Scare and Cold War. A series of mid-twentieth-century events rang loud the world over, and Grant County did not escape the fright posed by an enemy who was understood to have threatened America with "We will bury you."[11] In 1950 the Korean War had just begun. China entered the war in support of North Korea, and China and the Union of Soviet Socialist Republics signed the Sino-Soviet Alliance. Over President Harry Truman's veto, the U.S. Congress passed the McCarran Internal Security Act, calling for severe restrictions against communism. Also known as the Subversive Activities Control Act of 1950—said by Truman to make a mockery of the Bill of Rights—the law contained emergency detention provisions, required members of the Communist Party to register with the government, prohibited communists from working for the federal government, and tightened undocumented exclusion and deportation laws. The passing of the Subversive Activities Control Act and the Empire mine strike occurred in the same year. It was wartime and a period of threatening worldwide communist expansion. These conditions provided an opportunity for giant mining corporations to reverse the employee gains made by unions during World War II. They initiated a coordinated, nationwide effort to seize the moment.

Grant County had not and likely will never again see such a long and tension-filled strike. On October 17, 1950, Local 890 called the strike against the Empire Zinc Company, and work stood still for eight months, until the following year. In June Empire Zinc obtained an injunction outlawing picketing by union members, a move that appeared likely to lead to the displacement of strikers and pave the way for a new labor force. This was a turning point in the strike. The Mine-Mill Local 890 meeting on June 12, 1951, presented miners with a daunting quandary. Not only was the local Anglo police power structure under Empire's influence, but now a local judge had stepped in with a court order that seemingly neutralized the union's most powerful tool—the right to strike and picket. The injunction meant striking union members who remained on the picket line would be jailed. The meeting ran into the early morning, teeming with raucous and unyielding disagreement.

Unknown to others at the meeting, Clinton's wife, Virginia, Virginia Chacón, and Aurora Chavez had met earlier in the day to figure out a way for the women to help. Virginia Jencks was bright and aggressive and seen by many as the bolder of the husband-and-wife duo, especially when it came to women's questions. The civil rights movement was a dozen years in the future, and the women's movement remained two decades distant. Although women had won the nationwide right to vote with adoption of the Nineteenth Amendment in 1920, remnants of an earlier era remained deeply entrenched in Grant County more than thirty years later. A woman had not yet served on a Grant County jury. Prostitution was an open secret that remained tolerated by local government, law enforcement, and Sheriff McDonald until Silver City's last brothel—memorialized by Max Evan's *Madam Millie: Bordellos from Silver City to Ketchikan*—on Hudson Street closed in 1968.[12] As with so many towns in 1950s rural America, social progress remained in Silver City's future. To the extent it was even known, it was an ideal, not yet a custom.

In a national setting where segregated schools were common and fear of a communist takeover pervasive, Clinton Jencks, the idealistic Anglo union organizer for Local 890, and his wife, Virginia, had joined the struggle. He was fiercely loyal to the union cause and remained so fifty years later—appearing frail, yet fully in control of his mental faculties—when he took part in the symposium commemorating the Empire mine strike and the 1954 movie in which he and most others played themselves. His account of the miners' and their families' response to the injunction against picketing revealed an insight which no one involved in these events could have possibly imagined. A history-altering change was in the making, well before it was formally recognized years later as feminism.

Recounting events, Jencks remarked, there was no formula to guide them; rather, they were making things up as they went along. He described how the union meeting dragged on with no solution in sight, when, seemingly out of nowhere, Teresa Vidal raised her hand and was given the floor. The movie may have embellished this scene, but it captures her proposal, the macho response, and its implications:

> Brother Chairman, if you read the court injunction carefully, you will see that it only prohibits striking miners. We women are not striking miners, we will take over your picket line. [Men laughing] Don't laugh. We have a solution, you have none. Brother Quintero was right when he said we will lose 50 years of gains if we lose this strike . . . your wives and children, too. But this we promise: if women take your place on the picket lines, the strike will not be broken, and no scabs will take your place.[13]

There was more debate and several votes but no decision. The union meeting was then changed into a community meeting, allowing the women's voices to not only be heard but their votes to be tallied. Finally, a motion for the women to staff the strike line passed by a slim majority. Although this was a small

FIGURE 38 Henrietta Williams as Teresa Vidal, addressing the critical union meeting. Publicity still, *Salt of the Earth*, ca. 1954. Collections of the Wisconsin Center for Film and Theater Research.

group in a small town in the sparsely populated Southwest, the decision was a hint of things to come for American women and a seismic shift for these Mexican American families. In Jencks's words, this was just like a "thousand other struggles" until the women took over the picket line.[14] Suddenly something remarkable had happened. As pointed out by Jencks:

> One, it was an all-male industry. None of these women were working actually on the job. . . . Their whole lives were involved, but who thinks about the women that are in the home. They're only struggling to stretch the paycheck. They're only doing, you know. *Only* . . . and that's the way we've been

> used to thinking . . . number two, against the background of a culture in which male dominance is exalted by the dominant religion, the Catholic Church. Machismo, you know. Holds the woman down, that whole thing. . . . So when this explosion comes . . . the women taking over is the thing that excited the imagination of people all over.[15]

Although this seemed to come unexpectedly, revolutionary ideas evolve incrementally. During the symposium, Jencks reminisced, "We all remember Sister Elvira Molano, who had lost one husband to silicosis in the mines and was losing a second . . . and without saying anything to anybody or getting permission from any man . . . first she would just stand on the side of the picket line, knitting, not saying anything, but she was there . . . without anybody saying anything, she just slipped into the line and started walking. . . . She didn't wait for the vote. Now that was working-class leadership when you know what's right."[16]

Thinking back on what turned out to be a historic moment and later became known as "consciousness raising," Jencks added, "The moment of consciousness is a sort of an academic term that describes that point that no one knows, but in actual life it's a process," implying there was no epiphany of a new consciousness to anyone present when Elvira Molano slid into the picket line or Teresa Vidal made the venturesome suggestion that women stand in for their husbands on the picket line.[17] Both events were peculiar. As culturally *prohibido* as it was for women to take over for men, there seemed to be no alternative. To the picket line the women went, the men assumed the household chores, and Empire Zinc would eventually throw in the towel.

Earlier efforts by Jencks to include the wives at union meetings had been flatly rejected by the men. The men had little interest of giving up control. Yet over time, their machismo crust softened, Jencks told the conference gathering. Initially, the men allowed the women to attend the post-meeting socials. Jencks spoke with a slight smirk when he recalled how, gradually, the wives found their way into the formal meetings, "and life was easier at home now because they were both on the same team."[18]

After serving on the picket line, the women returned to future meetings emboldened. Their thinking about how to contend with the opposition forces and their expanded position in the community were apparent from the strength of the women's voices. Strong talk was backed up with staying power when six women were charged with assault and battery. The tactic of the women taking control of the picket line led to a day-long battle that eventually resulted in massive arrests. Forty-five women, seventeen children, and a six-month-old baby were jailed, shocking observers, both locally and nationally. Numerous miners had been jailed, but this was different.

Local authorities appealed to the state government for help, but Governor Edwin L. Mechem refused a request for state police to serve as strike breakers. The women were then offered release if they would stay off the line. They refused, and local authorities caved. When the district attorney sauntered into the jail that evening and announced, "I'm going to take you *girls* home," they all shouted, "We're going straight to the picket line."[19]

Upon release, they chartered a bus and did just that. At this point, *Time-Life*, the *New York Times*, and even Silver City's *Daily Press* and *Silver City Enterprise* softened their stances and momentarily appeared to side with the strikers and their families.

Confrontations and confusion continued. But, in the face of the long-established mining industry's influence, the perceived threat of communism, and historic, racially based social pressure, there was only a modest shift in favor of the miners as the union's publicity remained ineffective. As the strike continued, opponents of the striking families began to speak of saboteurs, communists, and strikers in the same breath. The question of loyalty soon overshadowed notions of fairness when people discussed the strike.[20] Frequent clashes between women picketers and deputies, magistrate court cases, and arrest warrants were the order of the day. More violence followed on the picket line. The crisis escalated when hundreds of protesters filled the Grant County Courthouse, confining the newly elected sheriff, Leslie Goforth, to his office for several hours before state police cleared the crowd. Goforth had come into office as a result of Mexican American support and their opposition to McDonald.[21] At one point, Federal District Judge Archibald W. Marshall issued a contempt order, fined Local 890 and the international union four thousand dollars each, and sentenced Clinton Jencks and others to ninety days in jail. Jencks was forced to serve his sentence in solitary confinement. There was increasing picket-line violence: more attacks on women picketers, efforts to breach the picket line, rock throwing, gunfire, the use of vehicles as battering rams, approximately ninety charges for assault and unlawful assembly, and more.

FIGURE 39 Cowboys deputized by Sheriff Goforth and two women from the picket line tussle in one of the many confrontations during the fifteen-month strike. This photograph was used in Local 890 pamphlets during the walkout. Photograph, now lost, by union photographer José Carrillo.

Empire's intransigence has never been fully understood. Union negotiators repeatedly offered to accept any one of a number of contracts in force in other district zinc and lead mines. Empire uniformly refused. The National Labor Relations Board found Empire guilty of unfair labor practices for its refusal to bargain. After fifteen months, in January 1952, the twenty-fifth negotiating session in El Paso, Texas, finally paid off. The Empire Zinc Company and Local 890 settled. The strike was over. The miners secured a wage increase, additional vacation and pension benefits, improved housing conditions, improved safety standards, and other benefits. At long last, workers saw the end of the hated "Mexican wage." The resolution involved compromise, but the longest strike in the state's history was finally over, with more than a few union members injured but nobody killed.

Union members voted to decertify the seventy-two-year-old Local 890 in September 2014, yet its heritage remains. The union hall still stands in Bayard and bears a mural sequence created in 2005 by the Youth Mural Program of Silver City, commemorating the strike along the entire front of the building (fig. 40). An inscription below one section of the mural series quotes Howard Zinn: "But it did change. It changed because ordinary people organized and took risks and challenged the system and would not give up."[22]

Virginia Jencks's tough, stoic constitution was not unlike that of other women who had come to the Southwest before her. Virginia held to her personal belief in putting herself out front where people would have to take notice. In her words, "I live; therefore, I must defend others' rights to live." She embodied German philosopher Hegel's theory of moral action and his idea that no one is free so long as another is in chains. Mine-Mill union's Maurice Travis maintained that between Virginia and Clinton, she was the "greater trouble-maker of the two." Clinton Jencks's biographer James J. Lorence refers to Virginia Jencks as "a committed working-class feminist." He credits her as her husband's "conscience in the promotion of gender equity." She

FIGURE 40 Mural on union hall building, Bayard, New Mexico. Photograph by the author, 2020. Mural by the Youth Mural Program project, lead artist M. Fred Barraza, Cobre High School art students.

could also become belligerent, causing the Hispano women pause when Virginia suggested that protests would be effective if taken to town. Virginia's abrasiveness fostered Empire loyalists' view of the Jenckses as outside communist agitators who moved in to stir up trouble. *Salt of the Earth* director Paul Jarrico praised Virginia for being the "guiding force that inspired the feminism among the Latinas of Grant County."[23]

As with their pivotal participation in the Empire mine strike, women are playing an increasing role confronting the climate crisis and in environmental work today. Women occupy more than half of leadership positions in conservation and preservation organizations, comprise more than 60 percent of new hires and interns, and dominate executive directors' positions in environmental grantmaking foundations. The director and fifteen out of twenty-one board and staff members at the Quivira Coalition in New Mexico are women. Eunive Newton

Foote was the first American scientist to make the connection between carbon and the warming of the atmosphere in 1856 (more on Foote in chapter 11). The number of women scientists involved in today's climate change research and advocacy is in the thousands. Of course, Greta Thunberg's role inspiring the world's youth and alerting much of the planet's population to the plight of humanity not only speaks to the place of women, both children and adults, in the climate movement, but again confirms that one person can make a world of difference, and each of us can try.

The Empire Zinc Company and its supporters exploited the Jenckses' former association with communism as a tactical concoction to serve their opportunistic interests, but these claims do not stand up to Clinton Jencks's religious leanings and military service. His grandparents were missionaries, and his mother was "exacting in the standards demanded of her son," making clear "that a sense of Christian duty and moral responsibility was his birthright." Liberator bomber crew missions in Guam, where Jencks served as lead navigator on forty missions for a B-24 squadron during World War II, were highly perilous. For his achievements, Jencks was awarded six medals, including the prestigious Distinguished Flying Cross for "extraordinary heroism" and "high skill and courage" above and beyond the call of duty.[24]

The Empire mine strike had an afterlife, both personal and public. Although the rabid overkill and falsehoods of McCarthyism were long ago repudiated, historical revision is of little consequence to those workers branded by the spurious use of Red Scare tactics during the strike. Employment opportunities remained hard to come by for decades for those who were blackballed as a result. Speaking at the 2004 symposium, elderly union representative Albert Millan recalled that, during a 1967 strike, he traveled to California for work. Millan had been jailed and wrongly labeled a communist during the Empire strike, and the branding remained almost two decades later, forcing his return to New Mexico—blackballed. Another elder of the strike, Lorenzo Torrez, told the gathering that the FBI followed him until the mid-1980s, helping explain lingering Hispano resentment.[25]

After the strike and prompted by Jencks, several Hollywood filmmakers who were blacklisted in connection with the McCarthy hearings developed the now-famous movie idea and approached the families with a unique proposal. The filmmakers wanted to produce a motion picture that realistically provided an account of working-class families dismantling the racist dual-wage system. They proposed a first-ever artist-miner collaboration with equal and shared say in production decisions. The project suggested by the filmmakers was not only groundbreaking for its proposed collaboration, but it also initiated an inquiry into a broad swath of American life that had barely begun to emerge in the media: the social condition of Mexican Americans. *Salt of the Earth* is a landmark labor film, a "powerful cinematic expression of racial politics, class conflict and feminist discourse rooted in the struggle of Mexican American miners in the Southwest, men and women whose struggle for dignity had been magnified and distorted by the brutal persecutions of the McCarthy era."[26] The film won the International Grand Prize for best film in 1955, as well as being certified by the Library of Congress as one of only one hundred films ever produced to be preserved for posterity (see YouTube).

The final chapter of these events ended up in the United States Supreme Court after the criminal prosecution of Jencks in El Paso Federal District Court. Prosecutors claimed Jencks was a communist by his own admission. The jury trial resulted in a felony conviction based on false testimony, a verdict that was appealed and eventually reversed by the Supreme Court. It's the case that established an accused expanded Sixth Amendment rights. Despite Jencks's heroic role in serving his country in a time of war, his helping to bring about a just settlement of the Empire strike, enduring solitary confinement, several years of legal appeals, and his case serving as catalyst for the expansion of Sixth Amendment rights for every

American with the Jencks Act, Jencks was forced to give up his employment for the good of the union he so faithfully served and deeply admired. He retrained himself as a welder but was blackballed throughout the Southwest. He then turned to education, obtained a doctorate in economics, and taught at San Diego State University until his retirement in 1988.

The price that the miners paid for their involvement in the strike and the film is not well documented, but after the fifteen months of economic hardship, personal risk, and long-range consequences as demonstrated by Albert Millan's account of finding work in California, the miners were confronted with a revolving turnstile of hiring, firing, reinstatement, and being fired again. The *Salt of the Earth* film's leading lady from Mexico, Rosaura Revueltas, was deported before the film was completed and blacklisted in her country. Herbert Biberman's pivotal role in directing the film finally forced him out of the industry altogether and into selling real estate. Cast member Will Geer (who played Sheriff Goforth) was blackballed, though he later played Grandpa on *The Waltons*. Those who participated in the Empire mine strike and the filming of *Salt of the Earth* were grossly undercompensated and remain under-recognized.

Piecing together Bartley McDonald's role leading to the Empire strike, the film, the Supreme Court's *Jencks* case, and the relevance of this account to the Voice of the Streets has been gratifying. I wanted to know more, but by the time I made the connection between the seemingly disparate parts of this tale, it was too late to talk with Jencks: He died on December 14, 2005, at the age of eighty-seven. It's a disappointment to have missed the opportunity to know what he thinks about the Voice of the Streets and his one-punch encounter with the previous owner of this ranch that led to so much. Continuing tribute is due Clinton Jencks for his undying belief that "we can have individual dignity only when we all have dignity" and that while "one stick could be broken, many sticks bound together could not." Clinton Jencks will long be remembered, in the words of Marjorie Cohn, as a "true egalitarian."[27]

The legacy of the Jenckses, the Mine-Mill union members, the miners' wives, and the film makers provides an essential truth: when enough people have had enough of being ignored and wronged, their collective voices can rapidly bring about fundamental, structural change. Seismic paradigm shifts can occur when exceptional leaders, a thoughtful plan to replace the current system, bold, aggressive, unconventional action, and a committed citizenry on the front lines come together as the proverbial Voice of the Streets.

CHAPTER ELEVEN

Causes for Outrage

TO THE SURPRISE OF no one, money, greed, and ruthlessly cruel short-term thinking is at the heart of the Trifecta Crisis. What's far less known is how long scientists have been aware carbon in the atmosphere can warm the climate and how long the fossil-fuel industry, their collaborators, and the U.S. government have been gaming the system, science, their fellow human beings, and planetary life generally. These swindlers have been colluding in an immoral, conspiratorial, and fiscally driven collaboration that defies civic virtue and turns common sense on its head. They've soiled their own plate. If we are to survive, before this is over, the public will need to take to the streets, follow the lead of the brave women of the Empire Zinc Mine strike, and bring the corporate wrongdoers and their collaborators to their knees.

The recognition of carbon emissions as the cause of the greenhouse effect happened as one would expect of scientific discoveries. Pioneering findings about the atmosphere were recorded, written about, peer reviewed, replicated, and eventually made their way into the public sphere. Inexplicably, this knowledge was captured by the American brand of consolidated corporate capitalism, whose devotion to stockholder earnings and quarterly returns has led to short-term thinking and worsening crises. This way of thinking concealed the foreseeable risks of an overheated climate. Preoccupied with fostering truth, George Orwell wrote, "at any given moment there is an orthodoxy, a body of ideas which it is assumed that all right-thinking people will accept without question."[1] In our context, that orthodoxy is this: "Fossil-fuel-driven life is the way it is, regardless of the consequences."

Orwell's thinking may not fully explain how the confluence of convenience in our daily lives and disinformation about fossil fuel products vanquished science, but the industry and U.S. government, for a very *long moment* of more than half a century, were fully aware of the serious, inevitable risks if carbon emissions were not pulled back from the current 424 ppm. Yet somehow, we entered an equally long moment of near universal ignorance. The monied few didn't ask for a few coins from the working class; instead, they somehow secreted away the science and, with government cooperation, pilfered trillions of dollars from all of us and risked the future of everyone in the process.

The place to begin this account of the destruction of the environment and possibly even civilization is by reviewing just how long the greenhouse effect has been known. Those who discovered and wrote about it wanted to understand weather. They engaged in the scientific process of seeking truth and simply told it. Scientists have known about atmospheric carbon's potential to heat Earth and the greenhouse effect for nearly two hundred years. The overwhelming scientific certainty about Earth's warming, species extinction, and soil loss and the history of collapsed civilizations that failed to adjust to changes in their lands and climate make the long-standing failure of personal, corporate, and governmental response to these crises inexplicable.

The story of this human-caused climate catastrophe can be told as an account of carbon dioxide, heat, and wind that exist in the planet's three-hundred-mile-thick atmosphere. Understanding climate requires that we assess what goes on in the atmosphere, and that's what scientists have done. This review of atmospheric measurement begins in 1735, when a fairly unknown eighteenth-century Englishman by the name of George Hadley published "Concerning the Cause of the General Trade Winds" in the *Philosophical Transactions of the Royal Society*. Hadley discovered that trade winds were caused by tropical heat moving air toward the North and South Poles, then depositing it in the subtropics, from which it subsequently returned toward the equator.

Between Hadley's discovery and the late 1950s, a good deal was learned about climate and weather. Guy Callendar is credited with making the first diagrammatic observation of global warming in 1938, when he emphasized the influence of rising carbon dioxide on Earth's temperature in the *Quarterly Journal*

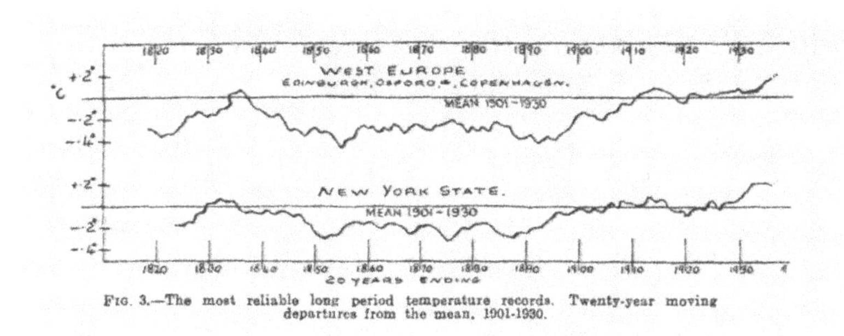

FIGURE 41 Diagram of rising temperatures in Western Europe and the state of New York between 1901 and 1930. Paper by Guy Callendar, figure 3, *Quarterly Journal of the Royal Meteorological Society*, 1938.

of the Royal Meteorological Society. Callendar's hand-drawn diagram describes a rise in global average temperature beginning about 1915—approximately forty years after the early spike in fossil fuel consumption.

A quote from Callendar's abstract explains the issue that worried him:

> By fuel combustion man has added about 150,000 million tons of carbon dioxide to the air during the past half century. The author estimates from the best available data that approximately three quarters of this has remained in the atmosphere. . . . The increase in mean temperature, due to the artificial production of carbon dioxide, is estimated to be at an average rate of 0.003°C. per year at the present time [and] world temperatures have actually increased at an average rate of 0.005°C. during the past half century.[2]

If these crises were not threatening civilization, it might be laughable that Callendar's observations predate the widespread indifference we've encountered for more than three-quarters of a century. He calculated, diagrammed, and explained the idea that humans, often thought of individually as mere grains of sand on the beach of life, could fundamentally change the world. Callendar appreciated the irony: "Few of those familiar with the natural heat exchanges of the atmosphere, which go into the making of our climates and weather, would be prepared to admit that the activities of man could have any influence upon phenomena of so vast a scale [but] I hope to show that such influence is not only possible, but is actually occurring at the present time."[3]

Interest in the implications of atmospheric carbon on weather and climate gained additional importance in the 1960s, when the young Scripps Institute of Technology student David Kelly made two discoveries while measuring carbon dioxide in the atmosphere: (1) there was less carbon dioxide in the atmosphere during the summer, when growing plants extract it in greater quantity than in winter; and (2) more importantly, atmospheric levels of carbon dioxide were rising year by year. The yearly increase in atmospheric carbon that Kelly documented confirmed with data in real time what a variety of scientists had known since the late 1700s and had been warning about since the early 1900s when telephone inventor Alexander Graham Bell gave a lecture about it. It's clear from a number of publications that the risk of atmospheric overheating has been known for more than a dozen generations (table 2).

The warnings laid out in table 2 were not limited to scientific writings: "Pollution of the Atmosphere," Nature, December 7, 1882 (citing Tyndall's 1863 research); "The Atmosphere," New York Times, January 6, 1883 (also relying on Tyndall's research and the 1882 Nature article); "Remarkable Weather of 1911: The Effect of the Combustion of Coal on the Climate—What Scientists Predict for the Future," Popular Mechanics, March 1912 (citing Arrhenius's 1896 research); and "Coal Consumption Affecting Climate," Braidwood Dispatch and Mining Journal, July 17, 1912, Australia. Many other articles were published between 1883 and 1912 in various newspapers, including the Philadelphia Inquirer, Kansas City Star, and York Daily. Coverage continued thereafter: "Warmer World," Time, January 2, 1939; "Is the World Getting Warmer?" Saturday Evening

TABLE 2 Scientific writings and warnings on greenhouse gases.

YEAR	PERSON	EVENT/PUBLICATION	CONCLUSION
Late 1700s	Horace Benedict de Saussure	Four glass boxes experiment	Swiss naturalist trapped solar energy, as does the atmosphere.
1824	Joseph Fourier	"Remarques générales sur les temperatures du globe terrestre et des espaces planétaires," *Annales de chimie et de physique*	French physicist argued for and coined the term *greenhouse effect.*
1814–29	Alexander Von Humboldt	*Personal Narrative of Travels to the Equinoctial Regions of the New Continent, During the Years 1799–1804*, vol. 4, 43–44	German geographer posited that deforestation was changing climate.
1827	Claude Pouillet	Experiments and papers	French physicist made findings supporting greenhouse effect.
1847	George Perkins Marsh	Lecture to the Agricultural Society of Rutland County, Vermont	American naturalist predicted human-caused climate change would result from burning fossil fuel.
1856	Eunice Newton Foote	"Circumstances Affecting the Heat of Sun Rays," paper read for the American Association for the Advancement of Science	"An atmosphere of [carbon dioxide] would give to our earth a high temperature."
1863	John Tyndall	Lecture: "On Radiation through the Earth's Atmosphere"	Irish physicist explained the greenhouse effect in a public lecture.
1896	Svante Arrhenius	"On the Influence of Carbonic Acid in the Air upon the Temperature of the Ground," *Philosophical Magazine and Journal of Science* 41	Swedish scientist calculated how changes in carbon dioxide could alter the temperature of Earth's surface—first person to use the term *greenhouse gases.*

TABLE 2 *continued*

1917	Alexander Graham Bell	*National Geographic Magazine*	American inventor predicted unchecked burning of fossil fuels would cause a greenhouse to become a hothouse.
1938	Guy Callendar	"The Artificial Production of Carbon Dioxide and Its Influence on Temperature," *Quarterly Journal of the Royal Meteorological Society*	English engineer charted the rise in global temperature beginning in 1915.
1957	Roger Revelle and Hans E. Suess	"Carbon Dioxide Exchange Between Atmosphere and Ocean and the Question of an Increase of Atmospheric CO_2 During the Past Decades," *Tellus* 9	Oceanographers called releasing carbon into the atmosphere an "experiment with the environment."
1965	National Science Foundation	President's Science Advisory Committee Report on Atmospheric Carbon Dioxide	Scientists warned President Johnson about climate change.
1972	J. S. Sawyer	"Man-Made Carbon Dioxide and the 'Greenhouse' Effect," *Nature* 239	American scientist predicted carbon dioxide increase of 25 percent and temperature increase of 0.6 degrees Celsius by 2000.
1988	James Hansen	Testimony before U.S. Senate	NASA climate expert gave the first congressional testimony about global warming.
1989	Bill McKibben	*The End of Nature*	McKibben's book warned the public about the threat of global warming.
2008	James Hansen	Testimony before U.S. Congress	NASA climate change expert testified before Congress a second time, and implored them to enact legislation to stop burning coal.

Post, July 1, 1950; "The Weather Is Really Changing," New York Times Magazine, July 12, 1953; "One Big Greenhouse," Time, May 28, 1956; "Ominous Changes in the World's Weather," Fortune, February 1974; and "Is Energy Use Overheating the World?" U.S. News and World Report, July 25, 1977.

These magazine articles are merely the tip of the iceberg. Spencer R. Weart's "hypertext history of how scientists came to (partly) understand what people are doing to cause climate change" lists approximately 1,780 magazine articles, science reports, papers, and studies between 1784 and 2001, thirty-six of these before 1900.[4] More than a century ago, telephone inventor Alexander Graham Bell could not have been clearer when he predicted "the unchecked burning of fossil fuels would have a sort of greenhouse effect. . . . The net result is the greenhouse becomes a sort of hot-house."[5] Yet the fate of humanity and other life on the planet has long been cast to the wind.

Slowly leaking from under the dark cover of corporate wrongdoing, damaging evidence has come to light about the details of the fossil fuel industry's massive cover-up of the science on the greenhouse effect. A Union of Concerned Scientists report revealed ExxonMobil spent $16 million between 1998 and 2005 on a network of more than forty front groups to discredit mainstream climate science. A later Greenpeace report revealed billionaire industrialists Charles and David Koch spent significantly more than Exxon with virtually the same groups. The Union of Concerned Scientists revealed ExxonMobil, BP, Chevron, ConocoPhillips, coal giant Peabody Energy Corporation, and Royal Dutch Shell were fully aware, for at least three decades, of the debilitating worldwide consequences of climate collapse but continued to spend tens of millions of dollars to promote arguments they knew to be false. A leaked American Petroleum Institute communications team's campaign memo laid out a plan largely patterned after the tobacco industry's deceptive strategy, which was famously encapsulated in an internal memo asserting that "doubt is our product."[6] Documents released by Greenpeace in early 2021 establish the fossil-fuel industry knew as early as 1967 of the risk of the crisis we're now suffering. The fossil-fuel public relations campaign is the most well-funded in history and has created a narrative of doubt. Despite disastrous weather the world over, don't for a second think the fraud of corporate denialism is over; it's alive and well, as indicated by the Heartland Institute's October 15–17, 2021, climate change conference at Caesars Palace in Las Vegas on the theme of "The Great Reset: Climate Realism vs. Climate Socialism." The announcement and registration website read: "The global climate agenda, as promoted by the United Nations, is to overhaul the entire global economy, usher in socialism, and forever transform society as one in which individual liberty and economic freedom are crushed."[7] The early 2023 report that Exxon actually predicted decades ago precisely what has happened shouldn't surprise us.

James Gustave Speth is helping with the Our Children's Trust litigation (chapter 9). A report prepared in anticipation of his trial testimony has been reconfigured into a 2021 book, *They Knew: The U.S. Federal Government's Fifty-Year Role in Causing the Climate Crisis*—"the most compelling indictment yet written of the government's role in the climate crisis"—detailing the government's complicity in what is the cruelest and most destructive conspiracy in history.[8]

To whom do we direct outrage? The lineup is a fairly long one. First, the fossil-fuel industry conspired, profited, and lied. Next, the U.S. government collaborated with these swindlers and failed to protect the governed. Third are the scientists who sold out and wrote papers and reports, shilling for the fossil fuel industry. Fourth, related fossil fuel industries and companies went along with the scam. Fifth, some responsibility rests with the scientists who studied, wrote papers, yet remained in their offices rather than flood the Lions Club, Optimist Club, and other civic groups to explain to the public what they knew and what lies were being told. Finally, each of us bears responsibility to limit our consumption and rethink our convenient way of life. And how about the media? The fourth estate is tasked with informing us whether the other three are doing good work or not. Where were they?

Suddenly, the dramatic breakdown in climatic norms—fire, storms, floods—makes denial less plausible, leading to a public relations shift from hard denial to a newer, softer strategy of deception. Deflection is the most common tactic: "the problem is your fault," we're told. BP introduced the world's first individual carbon footprint calculator, deflecting blame thus: "Fossil fuels didn't cause the climate crisis; your use of petroleum products did." Or "Don't blame us. Act responsibly. You're unwilling to change your lifestyle." The companies are "carbon shaming" the public and maintain they did not create the demand for fossil fuels, rather simply responded to consumer demand. The notion of "consumer sovereignty" is a fundamental early economics idea discredited by economist John Kenneth Galbraith with his immensely influential 1967 book, *The New Industrial State*. The fossil fuel industry's antiquated claim maintains that without consumer demand, there would be no fossil fuels. However, current economic thinking holds that large, consolidated, unregulated corporations use advertising and other tactics to flip the buy-sell equation from consumer sovereignty to producer sovereignty. The corporation's gimmick is to blame the consumer lifestyle. While there is truth to faulting consumption, the distinguishing detail is the industry knew atmospheric carbon was dangerous and concealed this science-based fact, then denied or disputed it when the danger became known, and now endeavors to flip the fault.

"Doomism," the "too-late" narrative, is deflection's companion in this latest public relations onslaught. If you see the problem as inevitable, you'll disengage, give up, do nothing different. Denial, diversion, deflection, and now doomism tactics delay the impetus to change the system, most importantly to eliminate fossil fuels. In the whole of American history, there has never been a more glaring and unconscionable demonstration of profit over anything and everything.

Despite the fossil fuel industry's early knowledge of the danger of fossil fuels, the misinformation scheme was implemented by the Global Climate Coalition—a group of fifty U.S. corporations and trade groups created the year after James Hansen warned Congress of the threat of global warming—to discredit climate science despite its knowing full well, by its own studies, that "the scientific basis for the Greenhouse Effect and the potential impact of human emissions of greenhouse gases such as CO_2 on climate is well established and cannot be denied."[9] A European journalist uncovered a confidential 1988 report titled "The Greenhouse Effect" in which oil giant Royal Dutch Shell acknowledged that planetary warming largely driven by fossil fuels could cause sea-level rise and changes in precipitation patterns, ocean currents, regional temperatures, and weather that could have major social, economic, and political consequences. The report even warned of the risk of litigation against the U.S. government and fossil fuel companies and the risk of "vigilante environmentalists" similar to those of the earlier fiercely anti-tobacco movement.

The coal industry also knew of the risks of increasing carbon dioxide. In an August 1966 article titled "Air Pollution and the Coal Industry" in the industrial publication *Mining Congress Journal*, James R. Garvey, president of Bituminous Coal Research, Inc., wrote,

> There is evidence that the amount of carbon dioxide in the earth's atmosphere is increasing rapidly as a result of the combustion of fossil fuels. If the future rate of increase continues as it is at the present, it has been predicted that, because the carbon dioxide envelope reduces radiation, the temperature of the Earth's atmosphere will increase and that vast changes in the climates of the Earth will result. Such changes in temperature will cause melting of the polar icecaps, which, in turn, would result in the inundation of many coastal cities, including New York and London.[10]

The fossil fuel industry adopted and has been pursuing the same game plan that the tobacco industry used to obscure the link between smoking and cancer. Many of the groups and affiliated scientists who have been vouching for the fossil-fuel polluters previously shilled for the tobacco industry. I'll not let go of these paired science-denying crimes: the strategy of deception that ignored the warnings of John Wesley Powell when the West was being flooded with newcomers and James Hansen's warning about the climate crises more than three decades ago. These like-kind strategies of ignoring, demeaning, and misrepresenting science were first advanced by railroad corporations, developers, and their lackeys in Congress to repopulate the West, were then adopted by tobacco companies to conceal the risk of cancer from smoking, and are now being recycled by Big Oil, Big Coal, and assorted conspirators to create doubt about the connection between fossil fuels and the climate catastrophe. What this country saw when so much of its population moved West and when smoking was promoted, the country is now seeing with fossil fuels. It's a conspiracy of incalculable theft and destruction. The mantras are well documented: railroads, "rain follows the plow"; tobacco, "doubt is our product"; and fossil fuel, "climate change is a myth." The dollar-an-acre five-year scheme, in which a homesteader could acquire 160 acres of land if they lived on it and improved it, and similar efforts to settle the West lasted nearly eight decades; tobacco's version lasted nearly as long; and the fossil-fuel industry's version, at more than fifty years as of this writing, should have ended decades ago.

It's been more than half a century since the National Science Foundation warned President Lyndon B. Johnson about the warming climate in 1965. At this late juncture, global heating is an emergency, and we don't have decades to fiddle while the planet burns. There is a terrifying, emerging consensus that this is not merely a weather crisis but an existential crisis that will leave virtually the entire world vulnerable to both physical devastation and economic ruin. In 2020 about 1 percent of the world's population lived in "a barely livable hot zone. By 2070, that portion could go up to 19 percent."[11] It's been suggested that we're at the dawn of America's Great Climate Migration, but soon there'll be no livable America in which to migrate.

The details and extent of emerging climate litigation go beyond the scope of this discussion yet warrant mention. A report titled "Smoke and Fumes: The Legal and Evidentiary Basis for Holding Big Oil Accountable for the Climate Crisis," released by the Center for International Environmental Law, details how to hold Big Oil accountable for its decades of denial, obfuscation, outright lying, and attacks on climatologists. Even though Big Oil had its own science predicting the future harm that would be caused by the industry, along with the opportunity and legal and moral duty to avoid this crisis, industry leaders chose instead the criminal path—acts of climate barbarism. Climate liability is now on the table with litigation in the United States, Germany, France, the Philippines, Norway, Uganda, Colombia, Mexico, Canada, India, and, by the time you read this, likely elsewhere. As CIEL Staff Attorney Steven Feit points out, this history is clear: "The individual data points in this report, viewed in aggregate, tell the story of an industry with advanced knowledge and expertise about climate change. There is a reasonable public expectation that if there was a problem with their products, big oil would have been the first to know about it and would have had a responsibility to affirmatively warn the public. Instead, they actively sowed confusion and doubt."[12]

The founder of 350.org, Bill McKibben, likens Big Oil's wrongdoing to criminal activity: "We're used to talking about the 'crime of the century,' but this heist is unique in geological time—these companies have done their best to steal our future as a species on the planet—the greatest crime in human history."[13] McKibben and others in this fight to save life on the planet maintain the extra heat we trap on the atmosphere every day is equal to the heat from four hundred thousand bombs the size of the one the United States dropped on Hiroshima, Japan, in World War II. As a result, between 1990 and 2020, we have seen twenty of the hottest years recorded to date. The oceans broke their heating record for the sixth consecutive year in 2021, absorbing heat equivalent to 7 Hiroshima atomic bombs detonating each second, 24 hours a day, 365 days a year. The average carbon dioxide levels in 2021–2023 reached a new record, 424 ppm.

Methane emissions are approximately three times their rate three decades ago.

Not that more is needed to cause outrage, but there are those in high places who want to simply eliminate any potential governmental structure capable of implementing the changes necessary to curb cascading environmental collapse. Despite encouraging signs in 2021, 2022, and 2023 of both awareness and implementation of positive change, fierce commitments remain to the status quo—to do nothing. The United States has suffered a curious, unprecedented, and challenging time. Former president Donald Trump and Environmental Protection Agency administrator Scott Pruitt mocked the climate crisis as a hoax and didn't believe that climate breakdown was real; rather, they claimed it was a myth created by China. Not only did Trump announce his intent to withdraw the United States from the landmark Paris Agreement, but he also scaled back the Clean Power Plan designed to curb emissions from power plants and a host of other environmental safety regulations. Pruitt was openly hostile to the agency he was chosen to direct. Donald Trump's 2016 campaign manager and former White House chief strategist Steve Bannon made it clear that President Trump's cabinet picks were aimed at the "deconstruction of the administrative state," the bane of free-market devotees. "Deconstruction" is a euphemism for "elimination." Bannon, a pardonee of former president Trump, made a remark that should make us all shudder: "I'm a Leninist and Lenin wanted to destroy the state, and that's my goal too. I want to bring everything crashing down, and destroy all of today's establishment."[14]

Before the January 6, 2021, insurrection at the U.S. Capitol, Steve Bannon told President Trump: "You need to kill this [Biden] administration in its crib."[15] Their coup d'état attempt is proof Bannon's threats to "destroy" government have roots. The peaceful transfer of power that's served the U.S. democracy so well was shattered with Trump's refusal to accept the election outcome and saboteurs' efforts in response to that refusal. On September 29, 2021, Bannon met with hundreds of Republican faithfuls at the Capitol Hill Club at the invitation of Republican presidential appointees, urging them to be prepared to "reconfigure the government" when a Republican leader takes office. He told NBC News in an October 2, 2021, telephone interview: "If you're going to take over the administrative state and deconstruct it, then you have to have shock troops prepared to take over immediately. I gave 'em fire and brimstone," meaning there need to be "pre-trained teams ready to jump into federal agencies" as soon as the next Republican president takes office.[16] The following day, Bannon doubled down, saying Republicans were going to have twenty thousand shock troops to carry out the "deconstruction."[17]

Even though President Biden has pledged to cut the U.S. output of climate-warming carbon gases by at least 50 percent below 2005 levels by the end of the decade, the future of meaningful governmental climate action remains fuzzy. There are those who see Biden's announcement of the United States' Nationally Determined Contribution of carbon in his 2021 Earth Day remarks as deceptive, greenwashed, and a capitulation to polluters that does *not* align with what climate scientists say is necessary to prevent global heating. On the other hand, there are seventy-four million Americans supporting those who have made it clear they will oppose any effort to curb carbon emissions. It's apparent the United States, like the rest of the planet's polluting nations, has not yet reached a consensus to take action. The Russia-Ukraine war and the world's pull-back from Russian oil—after the West paid Russia more than $8 billion in the first two weeks of the war—may lead to a worrisome emphasis on use of the major cause of carbon dioxide emissions and a retreat from the transition to net-zero energy sources. At the same time, if the war leads to an energy crisis, new thinking about energy efficiency may spike, as some scientists believe efficiency is the largest, cheapest, safest, cleanest, and fastest way to address it.

Before we leave the record of the disgraceful lies of the fossil-fuel industry, the collaboration and failure of the U.S. government to respond to the history of science-based warnings of the approaching climate crisis, and insurrectionists waiting in the wings, it's instructive to keep in mind just how certain science is about climate change. As one biologist framed the issue in 2014:

> We've recorded all sorts of climate change shift in multiple areas. However, the scientific process is consistent. Every single individual study that has been done has gone through the same rigorous process, data collection, research analysis, and qualified peer review. At the moment, we've at least 10,000 different papers, completed over 20 years, each using different data sets, and they are all coming to the same climate change conclusions. We've a weight of evidence that the average person is simply not aware of—and this frightens me. I'd like to think that we're not going to reach the projected four degrees of warming this century because I can't even imagine what that would look like. [Eighty] years is not that long, and unless we act soon, my seven-year-old daughter will probably have to live through that.[18]

These findings, now more than fourteen thousand papers, are the result of the strict laws of science—without mercy and without exception.

CHAPTER TWELVE

The Voice of the Streets

TAKING TO THE STREETS is a formidable task in view of two obscure "somethings," both puzzling. First, we're burdened by our inherited addiction to convenience below the threshold of our conscious awareness. It conflicts with our taking action. Those of the Greatest Generation struggled but eventually enjoyed a world of plenty and the conveniences that followed. We've inherited their thinking. Susan Neiman's writing about hereditary racial bias has similar application to our inherited dependency on convenience, a barrier to street action: "None of us can entirely escape the residues of attitude transmitted from mother to daughter, father to son, unless we are bitterly scrupulous. Even then, those who make an effort to reject those attitudes are likely to retain traces."[1] We need to be scrupulous.

The companion to the flaw of unwittingly prioritizing convenience is an even more enigmatic failing: the "spectacle" discussed in chapter 9. The comfort and security of detached remoteness keeps us entertained and safe, yet disengaged and inactive. To become a fugitive from these social constructs—both are conceptual blinders—to abandon the inherited conveniences to which we've become accustomed, to step out of the spectacle, to foster the impulse to protect our children, and to become actively engaged residents and shapers committed to survival of planetary life is a full-time psychological occupation that will take steadfast effort.

If the push to accelerate solutions to climate breakdown, species extinction, soil loss, and other crises is to keep pace with the threat of environmental collapse, the streets need to be filled with protesters demanding an end to fossil fuel use, drilling for oil, oil pipelines, railroad transport lines, oil refineries, fracking, and the other ancillary enterprises tied to these carbon-spewing swindlers. It's past time for the U.S. Congress and state legislators to be overwhelmed with protesters demanding the end of subsidies and other legislative gratuities for corporate polluters and abuses by Big Oil, Big Ag, Big Pharma, Big Tech, and Big Banks. The Voice of the Streets can demand an end to the spoliation of the electoral process by the Big Donor class and the ruin of news and commerce by consolidated American corporations. Earth no longer has the capacity to support "bourgeois luxury." Our forebears initiated an experiment in governance designed to serve most if not all of us, rather than the privileged few who currently garner unimaginable wealth and obscene political influence. The Founding Fathers could not have imagined the deluge of such a vast amount of dark money into American political campaigns, the virulent social media, a politicized judiciary, nor such extreme wealth and income inequality.

While it's essential we promptly take action to solve this panoply of crises, working to end wealth and income disparity will best increase our chances of adopting solutions for the Trifecta Crisis. Far too many Americans are suffering under a rigged economic and political system. The institutionalized cheating in business, the consolidated brand of unregulated American corporate capitalism, and the dark-money-fueled politics benefiting the privileged few thwart initiative, progress, and solutions. For the first time in history, in 2018, $1.5 trillion tax cuts helped the richest four hundred families in the United States pay a lower tax rate than the working class: 23 percent for billionaires compared to 24.2 percent for wage earners. The tax rate for those richest few families was 56 percent in 1960 and has been falling ever since. The 2023 Oxfam study reports the wealth of the richest 1 percent spiked as two-thirds of new post-pandemic wealth was realized by this thin slice of the world's population, increasing their share of new wealth by $26 trillion. To be clear, 63 percent of the total new wealth went to the 1 percent, while 27 percent went to the remaining 99 percent of the world's population.[2] It's the same problem, the absence of regulation on consolidated corporations and regressive taxation. The report found 143 of 161 countries froze tax rates for the rich during the pandemic, and 11 countries actually reduced them, fostering this obscenely skewed post-pandemic profit boom. Pulitzer Prize–winner, ordained Presbyterian minister, journalist, and former war correspondent Chris Hedges argues that creeping racism in America calls for Reinhold Niebuhr's "sublime madness," our inherent capacity to defy the forces of repression and senselessness threatening civilization. Why? Hedges noted, "Elon Musk [and] Jeff Bezos . . . they're each worth about 180 billion dollars. . . . Now, when David Rockefeller died [in 2017], he was worth about 3 billion [dollars]. I mean, this is wealth amassing by individuals that we've never seen."[3]

Congressional intransigence has created the imperative for a comprehensive grassroots campaign to raise our voices above the din of contentment and unwitting collusion. This requires we leave the comfort of our homes, join with people we don't know and with whom we might otherwise disagree, and do things we've never done before. This must be a movement of ordinary people willing to march, take to the streets in mass, strike, boycott, submit to arrest and abrasive confrontation, invade (by stockholder election) corporate boardrooms, demand divestment, and push for commonsense reforms. Some of this action agenda will not be safe, comfortable, popular, or, in some instances, even legal, but in terms of entrenched customs, for the necessary many, these efforts are essential.

This is an abrupt pivot away from anthropocentrism—humans perceiving nature as a resource—and toward an ecocentric perspective and our constitutional right to be left alone, to not be deprived of our right to enjoy nature, to enjoy breathable air, to be free from catastrophic weather, and to live within a life-sustaining climate. You'll need to step outside the unquestioned norms of the dominant culture, a steep remove from social convention. Civil disobedience and direct action—climate crisis ambition, the decision to become an active part of the solution—are the key and the only means of transformative change when power is so deeply entrenched, as it is now in the United States. "The Voice of the Streets," street action, has been with us so long, even medical writers Rupa Marya and Raj Patel recognize that "great change has always come from incensed citizens ready to agitate for it."[4] Yet, so far, this most important moment in human history has been met with indifference.

The exasperation of the University of Edinburgh's Elizabeth Cripps regarding the world's apathetic response to what she sees as terrifying bears repeating: "It seems unbelievable that we're not all chaining ourselves to the headquarters of oil and gas companies, or at least hammering on MPs' office doors."[5]

There is enormous strength in social norms of business as usual, despite clear climate science and catastrophic climate disasters the world over. Yet ever more numbers of serious people are searching for ways to break through the mass delusion of normalcy, when so little is being accomplished to address this unique moment in human history, the accelerating breakdown of climate that is collapsing our life-support systems. Climate activists have upped their game with a barrage of tactics, like throwing tomato soup on a glass-covered Van Gogh's *Sunflowers* in the National Portrait Gallery in London in October 2022. In terms of press coverage, it was arguably one of the more effective climate actions to date—the tomato soup heard around the world.

The masses are equipped to speak truth to power, to act to ensure the future of the world's children and grandchildren, and to restructure how humankind lives on Earth. After writing *The Grapes of Wrath*, John Steinbeck famously said, "I want to put a tag of shame on the greedy bastards who are responsible for this." The institutional betrayal that has destabilized our global life-support systems provides the governed the right to feel and do the same. After all, this is the most important moment in the history of civilization, the potential collapse of our entire suit of life-support systems. The adage often attributed to Dante, accurately or not, is also worth repeating: "The hottest places in hell are reserved for those who in times of great moral crisis maintain their neutrality."[6]

Only 55 percent of eligible voters cast ballots for the president of the United States in 2016, a twenty-year low. That percentage was up to 66 percent in 2020, the highest in 120 years, yet still not quite two out of three eligible voters. An idea adopted by many, including Mario Vargas: "There is nothing that erodes and corrupts a political system as much as lack of popular participation, and that the responsibility for public matters should be confined—at the expense of everyone else—to a minority of professionals."[7] The political elite are standing in for the vast majority, and public matters entrusted to them are being resolved in keeping with the interest of only wealthy donors who finance the electoral success of those in Congress, science be damned. Referred to as "congressional stagnation," 98 percent of all incumbents are reelected. Gerrymandering and the antiquated Electoral College—this money-based skewing and stale, festering electoral system—have given us presidents who receive fewer votes than the opposing candidate.

The necessary transformative change needed to thwart the onslaught of global warming will require *hundreds of thousands* of committed risk-takers like Clinton and Virginia Jencks and the wives of union members who stood up to the powerful mining interests in the Empire mine strike; like the "Tank Man," also

FIGURE 42 Lucinda Cole (*lower right*), wearing a hat and with a butterfly tattoo on her outstretched arm at the September 11, 2011, White House protest against the Keystone XL Pipeline. Photographer unknown.

known as the "Unknown Protester," whose identity and fate remain a mystery but who defied a column of tanks in Tiananmen Square on June 4, 1989; the twenty-eight-year-old licensed practical nurse and mother who peacefully resisted during the 2016 Black Lives Matter protest in Baton Rouge, Louisiana; eight Jewish clergy members and activists arrested on Capitol Hill in the January 2018 street action protesting President Trump's plan to end the Deferred Action for Childhood Arrivals, the program protecting undocumented immigrants who came here as children; those protesting George Floyd's police murder; the more than one hundred thousand people who took to the streets of COP26 host city Glasgow, Scotland, to demand urgent, meaningful climate action; the four British activists found not guilty for knocking down and dumping the statue of slave trader Edward Colston and dumping it in the harbor of River Avon; and the millions of people who have taken to the streets to achieve the sixteen important social changes discussed in this chapter.

The energy transition from a fossil-fuel-based economy to renewables and a net-zero economy requires the extraordinary. It requires far more than the 1,253 arrestees who, along with James Hansen, actress Daryl Hannah, Sierra Club Executive Director Michael Bloom, civil rights veteran Julian Bond, and John F. Kennedy Jr. were cuffed, booked, briefly jailed, and fined $100 each in the September 11, 2011, Keystone XL Pipeline protest at the White House—one of the largest undertakings of civil disobedience and arrests in the history of the North American climate movement. The effort bore results. President Obama rejected the project. But soon after his election, President Trump allowed the pipeline to move forward. President Biden again yanked the permit. Incrementalism and can-kicking until tomorrow is suicidal. The Voice of the Streets must not only become louder, but the activism must become more dramatic in order to flood jails with arrests and courthouses with caseloads beyond capacity. The nation's streets, jails, courthouses, and White House staff are incapable of effectively contending with telephone calls, marches, arrests, trials, and nonstop protests of great magnitude. The immensity of the Trifecta Crisis makes civil disobedience a necessity and requires the limited capacities of U.S. jails and courts be aggressively exploited.

Overcoming the climate and companion crises also entails bypassing central government, focusing on divestment, blocking infrastructure development, engaging in massive nationwide civil disobedience, peacefully submitting to arrest, entering pleas of "not guilty," and demanding the use of the "necessity defense." Although not universal and uncommonly used, this defense holds that "an act is justified if it is by necessity taken in a reasonable belief that the harm or evil to be prevented by the act is greater than the harm caused by violating the criminal statute . . . [and is available when] the social pressure of circumstance causes the accused to take unlawful action to avoid a harm which social policy deems greater than the harm resulting from a violation of the law."[8]

These trials, whether before jury or judge, could overwhelm the judiciary. This overwhelm-the-system tactic is not an unrealistic or untested strategy. In 1910, when the Spokane city council passed an ordinance outlawing "soapboxing" or pro-union street-corner advocacy, the Industrial Workers of the World supporters overwhelmed the town—"soapboxing and goading the police to fill the jails until taxpayers cried uncle"—with the goal of overloading the oppressive system. After several months, protesters had so clogged the jails, created such a logjam in the courts, and cost Spokane taxpayers so much money the city council conceded defeat. The combination of passive resistance and direct action brought the government to its knees: jails were cleared of all free-speech defendants, and the city council repealed the anti-soapbox ordinance and made a number of other concessions.[9]

The climate change and companion crises prompt far more serious action than mere marching. Emily Johnston—arrested for shutting down a pipeline carrying Canadian tar sands crude oil into the United States—has made her thinking for the necessity of aggressive action clear: "We knew we were at risk for years in prison. But the nation needs to wake up *now* to what's coming our way if we don't reduce emissions boldly and fast; business as usual is now genocidal."[10] Although the outcomes in four related cases were mixed, the case against Johnston and co-defendants Annette Klapstein and Ben Joldersma was dismissed by Minnesota District Court Judge Robert Tiffany due to insufficient evidence that the "Valve Turners" damaged the pipeline. Despite the October 2018 victory, there was disappointment, noted by defendant Johnston: "While I'm very glad that the court acknowledged that we did not damage the pipelines, I'm heartbroken that the jury didn't get to hear our expert witnesses and their profoundly important warnings about the climate crisis. We are fast losing our window of opportunity to save ourselves and much of the beauty of this world. We turned those valves to disrupt the business-as-usual that we know is leading to catastrophe, and to send a strong message that might focus attention to the problem."[11]

Although an earlier appeals court ruled that the defendants could offer corroborating expert testimony by former NASA scientist James Hansen, 350.org cofounder Bill McKibben, Harvard Law professor Lawrence Lessig, and experts on political science, pipeline safety, and other topics in support of the necessity defense, the dismissal of the case short-circuited what activists and legal scholars hoped would be the first test of the "necessity defense" in climate crisis litigation. James Hansen remarked:

> It's great that the defendants were found not guilty, but we missed an opportunity to inform the public about the injustice of climate change. Now we need to go on offense against the real criminals, the government. The government, especially the Trump Administration, is guilty of not protecting the constitutional rights of young people. They should have a plan to phase down fossil fuel emissions, but instead, they aid and

abet the expansion of fossil fuel mining, which, if not stopped, will guarantee devastating consequences for young people.[12]

The time for our passive collaboration, complicity, and unwitting obedience has passed. The word of the year for 2017 was *complicit*, and most of us have been delinquent—by sleep-walking and living our lives of convenience the way we and our parents always have—even though confronted with these crises. Complicity is a core concern because of the extent to which we have deferred to and thereby sanctioned, approved, and actively supported—via rabid consumption—the behavior of powerful, consolidated, unregulated corporations. Wendell Berry has long concerned himself with this flaw, writing in 1998:

> What is not sufficiently clear, perhaps to any of us, is the extent of our *complicity*, as individuals and especially as individual consumers, in the behavior of corporations. . . . most people in our country, and apparently most people in the "developed" world have given proxies to the corporations to produce and provide all of [our] food, clothing, and shelter. . . . A change of heart or of values without a practice is only another pointless luxury of a passively consumptive way of life.[13]

Together, masses of people in the streets can accomplish what none of us can achieve on our own. What the Valve Turners and so many others like them are calling for is action against the pull of complicity, despair, passive resignation, indifference, life of the spectacle, and other conditions that cause us to stay in our culturally assigned places. The part we can play in the solution to these crises is to engage in it personally and right away. We needn't ever forget the three most essential words of the U.S. Constitution: "We the people." It's the people—not the states, not the federal government, not Congress, not corporate America—who hold ultimate power. The central power is with the people over government, not the other way around. That centrality is shifting to our youth:

"Young people are about to utterly transform climate politics."[14] As Greta Thunberg said, "Yes, we must vote, you must vote, but remember that voting only will not be enough. We must keep going into the streets."[15]

Had it not been for the Voice of the Streets, critical gains in American society would have been delayed or possibly would not have been made at all. In these issues or events, the power structure favored the status quo; conservatives opposed change; yet the Voice of the Streets prevailed:

- American Revolution
- Emancipation of slaves
- Overturning of Jim Crow laws
- Child labor laws
- Free grade school and high school education
- Minimum wage laws
- The eight-hour work day
- Social Security Act and Medicare
- Women's right to vote
- GI Bill of Rights
- Civil Rights Acts of 1957, 1964, and 1968
- Equal pay for women
- Right to abortion (momentarily trumped)
- 9/11 First Responders Bill
- Affordable Care Act
- Marriage equality (same-sex unions)

The Voice of the Streets is typified in the Standing Rock Sioux Tribe's Dakota Access Pipeline protest, where hundreds of participants from multiple tribes and their supporters—along with supporting protests in Atlanta, Georgia; Washington, D.C.; and Los Angeles, California—prompted the Obama administration to intervene and suspend construction of the 1,172-mile, $3.8 billion project.

In other actions in Europe, the government in Warsaw, Poland, promptly reversed its position on an anti-abortion law when thousands of Polish women dressed in black, went on a nationwide strike, and protested against a legislative proposal banning abortion. In France, President Emmanuel Marcon suspended a gasoline tax increase after weeks of Yellow Vests protests. In the United States, Pennsylvania nurse Ieshia Evans decided to join the call to stop police killings of Black men, left her five-year-old son at home, and ended up arrested after refusing to leave the motorway in front of the Baton Rouge, Louisiana, Police Department headquarters. She said later, "When the armored officers rushed at me, I had no fear. I wasn't afraid."[16] Attending her first protest, Evans abandoned the comforts of home to join the Voice of the Streets.

Taking part in the Voice of the Streets is not merely a function of youthful enthusiasm. Lucinda and I spent several hours with a woman arrested at the Keystone XL Pipeline protest after our release from jail (fig. 43). Like Ieshia Evans, she too was taking part in her first protest. This elderly woman left her home, knew no one, and was among people from different backgrounds and with whom she might disagree on any number of things. But she agreed with them about at least one thing: oil needs to stay in the ground. She was proud to have participated in the protest and her arrest as one of more than a thousand people at the White House, telling us she was "honored" and that this "meant something" to her.[17] We no longer recall her name but remember her children encouraged her to attend the protest and paid her way because they had young children at home, were working, and could not participate.

We need to stand alongside these women. The International Energy Agency warned in a June 2020 report that we had only six months in which to change the course of the climate crisis and prevent a post-lockdown rebound in greenhouse gas emissions that would swamp efforts to stave off climate catastrophe.[18] At this writing, we're nearly four years past that point. For the first time, turning a hefty corner, the agency's 2021 report makes clear: "There is no need for investment in new fossil fuel supply in our net zero pathway."[19] R. Buckminster

FIGURE 43 A protester being arrested in the Keystone XL Pipeline protest in 2011. Photographer unknown.

Fuller's admonition applies: "You never change things by fighting the existing reality. To change something, build a new model that makes the old model obsolete."[20] To reach the speed of emissions cuts necessary to save ourselves, we need to build the new on an unprecedented scale, in ways that change the fundamental workings of current systems, foreclose high emissions choices, and end the economics of pollution everywhere.

The COVID-19 pandemic is a stark warning of the risks posed by the Trifecta Crisis. It's clear the root cause of this pandemic is a function of humankind's disregard and exploitation of nature. The spillover of disease from animals to humans that likely caused the COVID-19 virus—as well as SARS, bird flu,

Ebola, MERS, Hendra, Marburg, Nipah, and HIV—is analogous to the overflow of greenhouse gases and other pollutants caused by our convenient and consumptive lifeway overheating the planet, extinguishing species, and eroding soil. The Harvard Global Health Institute and Harvard T. H. Chan School of Public Health hosted a group of experts from around the world to call attention to the overlooked link between planetary and public health. They point out pandemic prevention requires the end of deforestation, a goal having comparable application to the Trifecta Crisis. It turns out the COVID-19 pandemic caused a global recession and created soaring unemployment, but it also could have fostered a unique opportunity to boost economic growth, create millions of new jobs, put global greenhouse gas emissions into structural decline, and shape economic and energy infrastructure for decades to come. It was an opportunity to see whether the world might meet its long-term energy and climate goals. There are those who suggest the Russia-Ukraine War has created even more cause to rid ourselves of fossil fuels. Boosting the Russian economy and its war effort by purchasing the nation's oil has complicity implications. The International Energy Agency is the world's gold standard for energy analysis and established the first global blueprint for a green recovery from the climate crisis. Adhering to this plan will immediately solve economic decline and unemployment, as well as providing a chance to solve the climate and companion crises. The Voice of the Streets can pressure the U.S. government adopt the International Energy Agency's recommendations.

Coercing adoption of those recommendations is possible. There are hundreds of recent examples where people have taken to the streets, council chambers, and politicians' offices to pressure elected representatives and administrators to do the right thing. Near where Lucinda and I live in New Mexico, citizens joined the Voice of the Streets to protect Chaco Culture National Historical Park and the communities of Greater Chaco. In December 2017 a number of groups were able to put a stop to an oil and gas fracking measure that was thought to be failproof. A message from Miya King-Flaherty attests to our power: "The surprise outcome of the Dec. 14 Sandoval County Commission meeting gives us hope that when we take action against public policies that are bad for our health, our communities and our families, it truly matters. All your calling, emailing, testimony and rallying made a difference. Our collective efforts made this possible." The cover headline in the Sierra Club Rio Grande Chapter's newspaper read in bold letters: "United, We Win—public engagement—people showing up—won an astonishing series of victories, victories just when all seemed lost. But there's no time to rest."[21]

The protests that followed the police killing of George Floyd saw Americans taking to the streets and expressing rage born of despair. After decades of police brutalizing and killing Black Americans yet rarely facing consequences, Americans finally did what Americans always do when the status quo becomes intolerable to an overwhelming number. The murder of George Floyd triggered an extraordinary outpouring of sorrow and anger about police brutality toward Black Americans. That summer, more than 25 million people joined in protests in support of Black Lives Matter. We've recently seen the spoils of protest in Grant County after a half century of commercial efforts to dam or divert New Mexico's last free-flowing river, the Gila. In June 2021, the New Mexico Interstate Stream Commission finally conceded defeat and voted to end its efforts to dam the Gila River, the fourth time since the 1970s.[22] Citizen opposition was key. It always is. The voice and volume of collective action heard by the Sandoval County Commission and the Interstate Stream Commission were the Voice of the Streets.

Bulgarian writer Maria Popova maintains "critical thinking without hope is cynicism, but hope without critical thinking is naivety."[23] Hope is essential: a life-affirming antidote to the cowardliness of cynicism and a path out of the swamp of complicity, convenience, consumption, and spectacle. It's what allows us to resist what Barry Lopez called "the temptation to despair."[24] As used here; hope's actionable nature differs from the laziness of passive optimism. The goal of protest movements is to "provide hope and inspiration for collective action to build collective power to achieve collective transformation, rooted in grief and rage but pointed towards vision and dreams."[25]

We were heartened to see the success of Chaco drilling opponents, as we harbor an affection for Chaco Canyon. Lucinda and I spent a number of Sierra Club service trips in Chaco, living in tents and helping with archaeological surveys, core drilling overhead beams in the Great Houses, and providing other labor because of meager funding. It's helpful to keep the lesson in mind of the Mimbreño people who lived along the Pitchfork Ranch's reach of the Burro Ciénaga and those of Chaco Canyon, who warn us about what happens when societies pursue unrestricted growth and the short-sighted exploitation of natural resources and fail to plan for and adapt to recognizable changes in climate. Those First Peoples were forced to abandon their homes and disperse, and traditions were lost.

The sixteen fundamental societal gains listed in this chapter are proof street action achieves results. Not only do we need to get down on our hands and knees and invest labor in restoring soil and habitat generally, but we must "reimagine our future, to identify a different road than the one that the prophets of technological innovation, or global climate change itself, is offering us."[26] A major part of that future is becoming a regular participant in the Voice of the Streets, giving voice to what needs doing and to do our part. Turning again to Einstein: "Those who have the privilege to know have the duty to act."[27]

"To act," to join the Voice of the Streets, is to become a member of Parliament or Congress of the Streets. We have a friend who lives by this homegrown principle: "The world is run by those who show up."[28]

Conclusion

THE ESSENTIALS OF THESE environmental crises are captured in a conversation between a mother, who is a member of my generation, and her twenty-six-year-old daughter in Barbara Kingsolver's novel *Unsheltered.* The mother was taken aback when her daughter said she was "waiting" for today's leaders to check out and asked if she was really waiting "for people to die." The daughter replied,

> Yeah. To be honest. The guys in charge of everything right now are so old. They really are, Mom. Older than you. They figured out the meaning of life, I guess, in the nineteen fifties and sixties. When it looked like there would always be plenty of everything. And they're applying that to now. It's just so ridiculous. . . . The permafrost is melting. Millions of acres of it. . . . It's so, so scary. It's going to be fire and rain, Mom. Storms we can't deal with, so many people homeless. Not just homeless but placeless. Cities go underwater and then what? You can't shelter in place anymore when there isn't a *place.*[1]

The mother in Kingsolver's book, like Kingsolver and me, is a member of the generation that adopted a neoliberal ideology and accompanying environmental destruction that intensified in the early 1970s, locking us into the crises with which we are now grappling. There's a cruel irony in the importance of having a sense of place and not having a place to go. Only when my generation acknowledges itself as the Destructive Generation will we have the credibility to assist today's young people in dealing with the risks engulfing our planet. They have the potential to become the Restoration Generation, saddled as they are with the mess we've left them. Borrowing from Thomas Paine's immortal words: these are the times that try our souls. The dispirited stay the course while the virtuous abandon their consumptive way of life and dig in with hope, imagination, and hard work to survive and save our overheated world.

Recall the comment in the introduction about the Greek term *krisis* (in English, "crisis"), the turning point of transition where the patient either recovers or dies. A similar term that refers to a different turning point is *bechira* in the Jewish spiritual tradition, meaning an "internal turning point" for the moment of choice between continuing to act out of habit or convention—think "convenience"—and knowing that we have the power to act in a way to truly benefit ourselves and others.

A bechira or choice point occurs at the moment when one can choose between a truth or life-affirming path or a false or life-denying path.[2] Choice points are absent when we act out of routine or convention. In the context of the climate and companion crises, the term *bechira* suggests we have the power to choose survival. The science tells us it's not too late to avoid planetary collapse. We have a choice; even though we're already ensnared in the Trifecta Crisis, we can choose to avoid the worst.

Albert Einstein said this long ago, and it matters now more than ever: "Try not to become a man of success but rather try to become a man of value."[3] We all want lives of direction and a time on Earth during which we live in a way that gives our lives meaning, purpose, and significance. Humankind has reached the point where each of us needs to tap into our capacity to transcend ourselves, to expand our imagination and empathy, to live in truth, and to create and pursue justice.

With this book, I hope to draw on that capacity and trigger the survival instinct in all of us to assume a part in the task of restoring our wounded world.

NOTES

Introduction

1. Germain Greer, *White Beech: The Rainforest Years* (New York: Bloomsbury, 2013), 1.
2. Wallace Stegner, *The Sense of Place* (Madison: Wisconsin Humanities Committee, 1986), 8.
3. Wendell Berry, *The World-Ending Fire: The Essential Wendell Berry*, ed. Paul Kingsnorth (Berkeley: Counterpoint, 2018), x.
4. Roy Scranton, *We're Doomed: Now What?* (New York: Soho Press, 2018), 3.
5. Bill McKibben, "The Next Decade Will Decide What the World Looks Like for Thousands of Decades to Come," *Huffington Post*, May 6, 2015.
6. Christian Parenti, *Tropic of Chaos: Climate Change and the New Geography of Violence* (New York: Nation, 2011), 5.
7. Fiona Harvey, "Humanity Is Waging War on Nature, Says UN Secretary General," *The Guardian*, December 1, 2020.
8. Dana Nuccitelli, "The New IPCC Report Includes—Get This, Good News," *Climate Science*, August 12, 2021; Helen Sullivan, "'Code Red for Humanity': What the Papers Say About the IPCC Report on the Climate Crisis," *The Guardian*, August 9, 2021; IPCC, "Summary for Policymakers," in *Climate Change 2021: The Physical Science Basis, Contribution of Working Group 1 to the Sixth Assessment Report of the Intergovernmental Panel on Climate Change*, ed. V. Masson Delmotte et al. (Boston: Cambridge University Press, 2021).
9. Brad Plumer and Raymond Zhong, "Climate Change Is Harming the Planet Faster than We Can Adapt, U.N. Warns," *New York Times*, February 28, 2022.

10. "Floods, Fires and Heat Waves: Michael Mann on 'The New Climate War' and the Fight to Take Back the Planet," *Democracy Now!*, July 16, 2021.
11. George Orwell, *Orwell on Truth*, ed. Adam Hochschild (Boston: Houghton, Mifflin, Harcourt, 2018), 154.

Chapter One

1. Greg Grandin, *The End of the Myth: From the Frontier to the Border Wall in the Mind of America* (New York: Metropolitan, 2019), 172.
2. Albert W. Thompson, *They Were Open Range Days: Annals of a Western Frontier* (Denver: World Press, 1946), 194.
3. Linda McDonald, as told to Victoria Tester, "Voice of a Ranch Woman: Living Through the Droughts," *Desert Exposure*, August 2008.
4. Jack Ramsey, "These Are More of My People," *Silver City Enterprise Press*, 1968, 11. Ramsey wrote an earlier book about local residents of note: *These Are My People* (Detroit: Harlo Press, 1966).
5. Patsy Adams, personal communication, 2014.
6. Bobby Sellers, personal communication, 2005.
7. Dana Fortenberry, personal conversation, May 2019.
8. Patsy Adams, personal communication, 2014.
9. Vicky Taylor, "Reminiscences: Mr. & Mrs. Jonnie McDonald," unpublished paper for Dale Giese, History of the Southwest, Western New Mexico University, August 6, 1971, Silver City Historical Society, Silver City, New Mexico.
10. Wallace Stegner, *The Sound of Mountain Water: The Changing American West* (Garden City: Doubleday, 1969), 19.
11. Mickey McDonald Lemon, "Taylor and Florence McDonald: Their Roots and Their Family," unpublished manuscript, 2005, 41, provided to the author courtesy of Mickey McDonald Lemon.
12. Patsy Adams, personal communication, 2009.
13. Patsy Adams, email, February 12, 2016.
14. Linda McDonald, as told to Victoria Tester, "Voice of a Ranch Woman: If You're Moving, You're Okay," *Desert Exposure*, July 2008, 2.
15. Robert Julyan, *The Place Names of New Mexico* (Albuquerque: University of New Mexico Press, 1996), 338.
16. John Mack Faragher, "North, South, and West: Sectional Controversies and the U.S.-Mexico Boundary Survey," in *Drawing the Borderline: Artist-Explorers of the U.S.-Mexico Boundary Survey*, ed. Dawn Hall (Albuquerque: Albuquerque Museum, 1996), 8.
17. Faragher, 9.
18. John Russell Bartlett, *Personal Narrative of Explorations and Incidents in Texas, New Mexico, California, Sonora and Chihuahua, 1850–1853*, vol. 1 (Chicago: Rio Grande Press, 1965), 303.
19. Bartlett, 303, 307. Emphasis in original.

20. Bartlett, 362.
21. Gray Sweeney, "Drawing Borders: Art and the Cultural Politics of the U.S.-Mexico Boundary Survey, 1850–1853," in *Drawing the Borderline: Artist-Explorers of the U.S.-Mexico Boundary Survey*, ed. Dawn Hall (Albuquerque: Albuquerque Museum, 1996), 48.
22. John C. Cremony, *Life Among the Apaches* (New York: A. Roman, 1868; Tucson: Arizona Silhouettes, 1951), 56.
23. Grandin, *The End of the Myth*, 110–11.
24. Christopher Ketcham, *This Land: How Cowboys, Capitalism, and Corruption Are Ruining the West* (New York: Viking, 2019), 23. It must be noted that this quote, with the term "settler," reflects the early American view that the West was open range, lacking occupants, when First Peoples had in fact inhabited the continent for millennia.
25. Lynn Jacobs, *Waste of the West: Public Lands Ranching* (Tucson: self-published, 1992), 10.
26. Jack Watson, *The Real American Cowboy* (New York: Schocken, 1985), 191.
27. Jacobs, *Waste of the West*, 10.
28. Watson, *The Real American Cowboy*, 49.
29. William deBuys, ed., *Seeing Things Whole: The Essential John Wesley Powell* (Washington, D.C.: Island Press, 2001), 1; John Wesley Powell, *The Arid Lands*, ed. Wallace Stegner (Lincoln: University of Nebraska Press, 2004), xvi.
30. Wallace Stegner, *Beyond the Hundredth Meridian: John Wesley Powell and the Second Opening of the West* (Boston: Houghton Mifflin, 1954), 359.
31. Stegner, 212.
32. Stegner, 229.
33. Stegner, 333.
34. David Gessner, *All the Wild That Remains: Edward Abbey, Wallace Stegner, and the American West* (New York: W.W. Norton, 2015) 9.
35. Watson, *The Real American Cowboy*, 191; Donald Worster, *Under Western Skies: Nature and History in the American West* (New York: Oxford University Press, 1992), 52.
36. Wallace Stegner, *The American West as Living Space* (Ann Arbor: University of Michigan Press, 1987), 17.
37. Leo Damrosch, *Tocqueville's Discovery of America* (New York: Farrar, Straus and Giroux, 2010), 138–39.
38. Watson, *The Real American Cowboy*, xvi.
39. Richard Hofstadter, *The Progressive Historians: Turner, Beard, Parrington* (New York: Alfred A. Knopf, 1968), 103–4.
40. Stegner, *The American West as Living Space*, 69; Wallace Stegner, *Where the Bluebird Sings to the Lemonade Springs: Living and Writing in the West* (New York: Random House, 1992), 103.
41. Edward Aveling, *An American Journey* (New York: Lovell, Gestefeld, 1885), 155.

42. S. Kara Naber, "Where Gila Meets History," *Desert Exposure*, June 2011, 2.
43. Julian Baca and Michael O'Donnell, *New Mexico Property Tax Study: Agricultural, Working and Natural Lands* (Prepared for the New Mexico Legislature, October 2021).
44. Rocky Mountain Climate Organization and National Resources Defense Council (NRDC), "Hotter and Drier: The West's Changed Climate" (New York: NRDC, 2008), http://www.nrdc.org/globalwarming/west/west.pdf.
45. Sid Goodloe, personal communication, 2005.
46. Yvette Paros, U.S. Fish and Wildlife Service, personal communication, February 9, 2013.
47. Libby Boone, personal communication, May 24, 2014.

Chapter Two

1. D. Gori, G. S. Bodner, K. Sartor, P. Warren, and S. Bassett, "Sky Island Grassland Assessment: Identifying and Evaluating Priority Grassland Landscapes for Conservation and Restoration in the Borderlands" (Santa Fe: The Nature Conservancy in New Mexico, 2012), 85.
2. *Cassell's Spanish Dictionary* (New York: Funk & Wagnalls, 1960, 2010).
3. Ellen Soles, email, October 7, 2022.
4. Ralph J. M. Temmink et al., "Recovering Wetland Biogeomorphic Feedbacks to Restore the World's Biotic Carbon Hotspots," *Science* 376, no. 6593 (May 2022), DOI: 1o.1126/science.abn1479.
5. Ella Jaz Kirk, "The Sustainable Future of Water," unpublished manuscript, sophomore project, March 26, 2014.
6. T. A. Minckley and A. Brunelle, "Paleohydrology and Growth of a Desert Ciénega," *Journal of Arid Environments* 69 (2007): 420.
7. David R. Montgomery, *Dirt: The Erosion of Civilizations* (Berkeley: University of California Press, 2007), 2.
8. "Hermits Peak-Calf Canyon Burned Area Emergency Response Phase 1 Assessment Report Summary," Forest Service, U.S. Department of Agriculture, Santa Fe National Forest, June 2022.
9. A.T. Cole and Cinda Cole, "An Overview of Aridland Ciénagas, with Proposals for Their Classification, Restoration, and Preservation," *New Mexico Botanist*, Special Issue, no. 4 (September 2015): 36.
10. Aldo Leopold, *The River of the Mother of God and Other Essays by Aldo Leopold*, ed. Susan L. Flader and J. Baird Callicott (Madison: University of Wisconsin Press, 1991), 184.
11. Ralph J. M. Temmink, Leon P. M. Lamers, Christine Angelini, et al., "Recovering Wetland Biogeomorphic Feedbacks to Restore the World's Biotic Hotspots," *Science* 376, no. 6593 (May 2022).
12. Leopold, *River of the Mother of God*, 191. Emphasis in original.

13. Thomas A. Minckley, Andrea Brunelle, and Dale Turner, "A Paleoenvironmental Framework for Understanding the Development, Stability and State-Changes of Cienegas in the American Deserts," in *Merging Science and Management in a Rapidly Changing World: Biodiversity and Management of the Madrean Archipelago III*, ed. G. J. Gottfried, P. F. Folliott, B. S. Gebow, L. G. Eskew, and L. C. Collins (Fort Collins: U.S. Department of Agriculture, Forest Service, Rocky Mountain Research Station, May 1–5, 2012), 82.
14. Bronson W. Griscom et al., "Natural Climate Solutions," *Proceedings of the National Academy of the Sciences* 114, no. 44 (October 17, 2017): 11645–50, https://doi.org/10.1073/pnas.1710465114.
15. Bob Alexander, *Six-Guns and Single-Jacks: A History of Silver City and Southwestern New Mexico* (Silver City: Gila, 2005), 49.
16. Richard Bradford, *Red Sky at Morning* (New York: J. B. Lippincott, 1968), 86.
17. Dean A. Hendrickson and W. L. Minckley, "Cienegas: Vanishing Climax Communities of the American Southwest," *Desert Plants*, 6, no. 3 (1984): 131.
18. Daniel McCool, *River Republic: The Fall and Rise of America's Rivers* (New York: Columbia University Press, 2012), 173–74.
19. Minckley et al., "A Paleoenvironmental Framework," 77.
20. Gregory McNamee, *Gila: The Life and Death of an American River* (New York: Orion, 1994), 80–81.
21. Stegner, *The Sound of Mountain Water*, 29.
22. Charles C. Mann, *1493: Uncovering the New World Columbus Created* (New York: Vintage, 2011), 62.
23. Cathryn Wild, "Beaver as a Climate Adaptation Tool: Concepts and Priority Sites in New Mexico" (Santa Fe: Seventh Generation Institute, May 2011), 4.
24. Richard Felger, email, September 23, 2012.
25. Ben Goldfarb, "A Biologist Figures Out How to Keep Beavers Alive on Western Landscapes," *High Country News*, Paonia, Colorado, November 9, 2015, 7 and 23.
26. William W. Dunmire, *New Mexico's Spanish Livestock Heritage: Four Centuries of Animals, Land, and People* (Albuquerque: University of New Mexico Press, 2013), 1.
27. Conrad Joseph Bahre, *A Legacy of Change: Historic Human Impact on Vegetation of the Arizona Borderlands* (Tucson: University of Arizona Press, 1991), 116.
28. Daniel G. Milchunas, "Responses of Plant Communities to Grazing in the Southwestern United States," General Technical Report RMRS-V GTR-169 (Fort Collins: U.S. Department of Agriculture, Forest Service, Rocky Mountain Research Station, 2006), 7.
29. Bahre, *A Legacy of Change*, 123.

30. Worster, *Under Western Skies*, 41.
31. U.S. Senate, *The Western Range*, Senate Document 199, 74th Congress, 2nd Session, Serial Number 10005 (Washington, D.C.: Government Printing Office, 1936), 3.
32. Bahre, *A Legacy of Change*, 111.
33. O. K. Davis, Tom Minckley, Tom Motoux, Tim Jull, and Bob Kalin, "The Transformation of Sonoran Desert Wetlands Following the Historic Decrease of Burning," *Journal of Arid Environments* 50 (2002): 410.
34. Davis et al., 411.
35. Minckley et al., "A Paleoenvironmental Framework," 82.
36. The destructive scouring process is discussed in a drone video of the ciénaga on Pitchfork Ranch: "Aridland Ciénagas," written and narrated by A.T. Cole, filmed and edited by Richard Helbock, 2014, available at "Burro Ciénaga," https://www.pitchforkranchnm.com.
37. T. A. Minckley, Dale S. Turner, and S. R. Weinstein, "The Relevance of Wetland Conservation in Arid Regions," *Journal of Arid Environments* (2013): 219.
38. Barbara Tallman, Deborah M. Finch, Carl Edminster, and Robert Hamre, eds., "The Future of Arid Grasslands: Identifying Issues, Seeking Solutions," Proceedings RMRS-P-3 (Fort Collins: Rocky Mountain Research Station, U.S. Department of Agriculture, 1998), https://www.fs.usda.gov/research/treesearch/39794.
39. Various scholars, scientists, groups and students pursue a variety of studies and surveys at the Pitchfork Ranch. Ongoing lists of species are maintained: Birds and butterflies, Dr. Dale A. Zimmerman, professor emeritus, Western New Mexico University; mammals and amphibians, reptiles and fish, Dr. Randy Jennings, professor emeritus, Western New Mexico University; moths, Dr. Clifford D. Ferris, professor emeritus, University of Wyoming; and plants and grasses, Grant County Native Plant Society and Kirsten B. Romig, botanist, U.S. Department of Agriculture, ARS Jornada Experimental Range. The inventories are available on the ranch website: www.pitchforkranchnm.com.
40. IPCC, "Climate Change and the Land," Special Report, 2019, https://www.ipcc.ch/srccl/.
41. Leopold, *River of the Mother of God*, 191.
42. Richard O. Prum, *The Evolution of Beauty: How Darwin's Forgotten Theory of Mate Choice Shapes the Animal World* (New York: Doubleday, 2017), 12.
43. Charles Darwin, *The Life and Letters of Charles Darwin*, vol. 2, ed. Francis Darwin (1887; London: John Murray Press, 1911), 90–91.
44. Prum, *Evolution of Beauty*, 11, 13.
45. J. Baird Callicott, ed., *Companion to* A Sand County Almanac*: Interpretive and Critical Essays* (Madison: University of Wisconsin Press, 1987), 157.
46. Callicott, 168.

47. Aldo Leopold, *The Essential Aldo Leopold: Quotations and Commentaries*, ed. Curt Meine and Richard L. Knight (Madison: University of Wisconsin Press, 1999), 161–62. Emphasis in original.
48. Laura M. Norman, email, January 26, 2023.

Chapter Three

1. Steve Leonard and Wayne Elmore, "Cows and Creeks: Can They Get Along?" in *Ranching West of the 100th Meridian: Culture, Ecology, and Economics*, ed. Richard L. Knight, Wendell C. Gilgert, and Ed Marston (Washington, D.C.: Island Press, 2002), 145.
2. Nathan F. Sayer, Ryan R. J. McAllister, Brandon T. Bestelmeyer, Mark Moritz, and Matthew D. Turner, "Earth Stewardship of Rangelands: Coping with Ecological, Economic, and Political Marginality," *Frontiers in Ecology and the Environment* 11 (2013): 348; Matthew R. Sloggy, Stefan Anderes, and José J. Sánchez, "Economic Effects of Federal Grazing Programs," *Rangeland Ecology and Management* 88 (May 2023): 1–11; Virginia A. Kowal, Sharon M. Jones, Felicia Keesing, Brian F. Allan, Jennifer M. Schieltz, and Rebecca Chaplin-Kramer, "A Coupled Forage-Grazer Model Predicts Viability of Livestock Production and Wildlife Habitat at the Regional Scale," *Scientific Reports* 9 (December 30, 2019); Gidon Eshel, Alon Shepon, Tamar Makov, and Ron Milo, "Land, Irrigation Water, Greenhouse Gas, and Reactive Nitrogen Burdens off Meat, Eggs, and Diary Production in the United States," *Proceedings of the National Academy of Sciences*, July 21, 2014.
3. Marco Springmann et al., "Options for Keeping the Food System Within Environmental Limits," *Nature* 562 (October 2018): 501–2.
4. George Monbiot, "Embrace What May Be the Most Important Green Technology Ever. It Could Save Us All," *The Guardian*, November 22, 2022.
5. Beef alternatives include Beyond Meat, Boca, Morning Star; chicken alternatives include Wholesome, Just Like Chicken; lamb alternatives include Gourmandelle; milk alternatives include Blue Diamond, Almond Breeze, Rice Dream, Silk; cheese alternatives include Miyoko's, Daiya, Chao, Violife, Treeline, So Delicious; cream alternatives include Califia Farms, So Delicious; and ice cream alternatives include Ben & Jerry's.
6. "Profitable Cattle Marketing for the Cow-Calf Producer," University of Georgia Extension Bulletin 1078.
7. Michael Kades, "Protecting Livestock Producers and Chicken Growers," Washington Center for Equitable Growth, May 5, 2022.
8. Sharon Guynup, "Address Risky Human Activities Now or Face New Pandemics, Scientists Warn" *Mongabay*, August 3, 2021.
9. Richard L. Knight and Courtney White, eds., *Conservation for a New Generation: Redefining Natural Resources Management* (Washington: Island Press, 2009), 196.

10. John E. Mitchell, "Rangeland Resource Trends in the United States: A Technical Document Supporting the 2000 USDA Forest Service RPA Assessment," General Technical Report, RMRS-GTR-68, US Department of Agriculture, Forest Service, Rocky Mountain Research Station, Fort Collins, 2000, 25, https://doi.org/10.2737/RMRS-GTR-68.
11. Andrew J. Plumptre et al., "Where Might We Find Ecologically Intact Communities?" *Frontiers in Forests and Global Change*, April 15, 2021.
12. Andrew J. Hansen, Richard L. Knight, John M. Marzluff, Scott Powell, Kathryn Brown, Patricia H. Gude, and Kingsford Jones, "Effects of Exurban Development on Biodiversity: Patterns, Mechanisms, and Research Needs," *Ecological Applications* 15, no. 6 (December 2005): 1893.
13. Sam Levin, "Silicon Valley Billionaire Loses Bid to Prevent Access to Public Land," *The Guardian*, August 10, 2017; Kathleen McLaughlin, "Class War in the American West: The Rich Landowners Blocking Access to Public Lands," *The Guardian*, January 20, 2018.
14. Linda M. Hasselstrom, "No Place Like Home," in *Ranching West of the 100th Meridian: Culture, Ecology, and Economics*, ed. Richard L. Knight, Wendell C. Gilgert, and Ed Marston (Washington, D.C.: Island Press, 2002), 51.
15. Edward O. Wilson, *Half-Earth: Our Planet's Fight for Life* (London: Liveright, 2016), 3.
16. "Investing in Conservation Generates Huge Returns for Economy, Study Finds," *CBC News*, July 8, 2020.
17. Paul F. Starrs, "Ranching: An Old Way of Life in the New West," in *Ranching West of the 100th Meridian: Culture, Ecology, and Economics*, ed. Richard L. Knight, Wendell C. Gilgert, and Ed Marston (Washington, D.C.: Island Press, 2002), 20.
18. Wallace Stegner and Page Stegner, *American Places* (New York: Crown, 1987), 216.
19. Richard L. Knight, "Field Report from the New American West," in *Wallace Stegner and the Continental Vision*, ed. Curt Meine (Washington, D.C.: Island Press, 1997), 183.
20. Leonard and Elmore, "Cows and Creeks," 145.
21. Sharman Apt Russell, *Kill the Cowboy: A Battle of Mythology in the New West* (Reading: Addison-Wesley, 1993), 5.
22. George Wuerthner and Mollie Matteson, *Welfare Ranching: The Subsidized Destruction of the American West* (Washington, D.C.: Island Press), 2002, xiii.
23. Andy Kerr, "Where's the Beef?" *Andy Kerr's Public Land's Blog RSS* (blog), August 13, 2022, http://www.andykerr.net/kerr-public-lands-blog/tag/social+cost+of+carbon.
24. M. Shahbandeh, "Total Number of Cattle and Calves in the U.S., 2001–2022," *Statista*, October 10, 2022; Olivia Rosane, "Humans and Big Ag Livestock Now Account for 96 Percent of Mammal Biomass," *EcoWatch*, March 23, 2018; Joe Loria, "Three Largest Meat Producers Rival Exxon in Gas Emis-

sions," *Salon*, December 8, 2017; David Griffith, "Does Cattle Livestock Contribute 51% of Human-Derived Greenhouse Emissions?" *Metafact*.
25. Wuerthner and Matteson, *Welfare Ranching*, xv.
26. Ketcham, *This Land*, 372.
27. "Sector at a Glance," USDA Economic Research Service, https://www.ers.usda.gov/topics/animal-products/cattle-beef/sector-at-a-glance.
28. Matthew N. Hayek and Rachael D. Garrett, "Nationwide Shift to Grass-Fed Beef Requires Larger Cattle Population," *Environmental Research Letters* 13 (2018): 084005.

Chapter Four

1. Christopher Norment, *Relicts of a Beautiful Sea: Survival, Extinction, and Conservation in a Desert World* (Chapel Hill: University of North Carolina Press, 2014), 82.

Chapter Five

1. William J. Ripple, Christopher Wolf, Thomas M. Newsome, Phoebe Barnard, and William R. Moomaw, "World Scientists' Warning of a Climate Emergency," BioScience 70, no. 1 (2020): 100.
2. Andrew Lee, "This Must Sound a Death Knell for Coal and Fossil Fuels, Before They Destroy Our Planet," *RECHARGE*, August 9, 2021; Fiona Harvey, "Major Climate Changes Inevitable and Irreversible—IPCC's Starkest Warning Yet," *The Guardian*, August 9, 2021.
3. "John Kerry Says Harrowing U.N. Climate Report Underscores 'Overwhelming Urgency' for Action," *Newsessentials Blog*, August 23, 2021.
4. Courtney White, *The Age of Consequences: A Chronicle of Concern and Hope* (Berkeley: Counterpoint, 2015), 3.
5. Jonathan Watts and Vaclav Smil, "Growth Must End, Our Economist Friends Don't Seem to Realize That," *The Guardian*, September 21, 2019.
6. Wilson, *Half-Earth*, 54.
7. David Wallace-Wells, *The Unthinkable Earth: Life after Warming* (New York: Tim Duggan, 2019), 3.
8. Doyle Rice, "Climate Change Could Lead to 'a Collapse of Our Civilization' According to Sir David Attenborough," *USA Today*, December 3, 2018.
9. David Graeber and David Wengrow, *The Dawn of Everything* (New York: Farrar, Straus and Giroux, 2021), 128.
10. Oliver Milman, "'This Is a Fossil Fuel War': Ukraine Climate Scientists Speaks Out," *The Guardian*, March 9, 2022.
11. Stefan Gössling and Andreas Humpe, "The Global Scale, Distribution and Growth of Aviation: Implications for Climate Change," *Global Environmental Change*, November 2020.

12. Chuck Collins, Omar Ocampo, and Kalena Thomhave, "High Flyers 2023: How Ultra-Rich Private Jet Travel Costs the Rest of Us and Burns Up the Planet," *Institute for Policy Studies*, May 2023.
13. Thomas Wiedmann, Manfred Lenzen, Lorenz T. Keyßer, and Julia K. Steinberger, "Scientists' Warning on Affluence," *Nature Communications* 11 (June 2020), https://doi.org/10.1038/s41467-020-16941-y.
14. Wiedmann et al., 3, 1, 7.
15. John F. Marra, *Hot Carbon: Carbon-14 and a Revolution in Science* (New York: Columbia University Press, 2019), 4.
16. Melinda Hemmelgran, "Jonathan Lundgren, Ph.D., Blue Dasher Farm and Ecdysis Foundation, Defines and Discusses Regenerative Agriculture," *Food Sleuth Radio*, August 27, 2021, https://foodsleuth.transistor.fm/episodes/jonathan-lundgren-ph-d-blue-dasher-farm-and-ecdysis-foundation-defines-and-discusses-regenerative-agriculture.
17. Quoted in Kate Sheppard, "Former Top NASA Scientist Predicts Catastrophic Rise in Sea Levels," *Huffington Post*, July 21, 2015.
18. "Species Extinction—The Facts," *IUCN Red List of Threatened Species*, https://springbrooknaturecenter.org/DocumentCenter/View/749/Species-Extinction-05-2007-PDF.
19. R. Dirzo, "Defaunation in the Anthropocene," *Science* 345, no. 6195 (July 25, 2014): 401–6, doi: 10.1126/science.1251817.
20. Norman Cousins, "History Is a Vast Early Warning System," *Saturday Review*, April 15, 1978.
21. Joseph J. Ellis, *American Dialogue: The Founders and Us* (New York: Alfred A. Knopf, 2018), 96.
22. John Wayne, BrainyQuote, https://www.brainyquote.com/quotes/john_wayne_382409.
23. Sean B. Carroll, *The Serengeti Rules: The Quest to Discover How Life Works and Why It Matters* (Princeton, NJ: Princeton University Press, 2016), 7 and 11.
24. Carroll, 8.
25. Carroll, 63. Emphasis in original.
26. James Lovelock, *The Ages of Gaia: A Biography of Our Living Earth* (New York: W.W. Norton, 1989), 15.
27. Lovelock, 3, 13.
28. Hannah Ritchie, "The State of the World's Elephant Population," *Our World Data*, December 1, 2022.
29. Montgomery, *Dirt*, 5.
30. "Soil Erosion Must Be Stopped 'to Save Our Future,' Says UN Agricultural Agency," UN News, https://news.un.org/en/story/2019/12/1052831.
31. Rupa Marya and Raj Patel, *Inflamed: Deep Medicine and the Anatomy of Injustice* (New York: Farrar, Straus and Giroux, 2021), 118.
32. Montgomery, *Dirt*, 16.

33. Patrick Greenfield, "Revealed: More Than 90% of Rainforest Carbon Offsets by Biggest Certifier Are Worthless, Analysis Shows," *The Guardian*, January 18, 2023.
34. Naomi Oreskes, "The Collapse of Western Civilization," *Living on Earth*, July 25, 2014.
35. Bruce Parker, "The Fourth Culture," *Worried About*, 247.
36. Tim O'Reilly, "The Rise of Anti-Intellectualism and the End of Progress," in *What Should We Be Worried About? Real Scenarios That Keep Scientists Up at Night*, ed. John Brockman (New York: Harper Perennial, 2014), 58.
37. Edward J. Watts, *Mortal Republic: How Rome Fell into Tyranny* (New York: Basic, 2018), 7, 8, 10.
38. For those reluctant to see the parallel between the Roman Empire and the United States and the repeated failure of civilizations, here is a survey of just how common it is that civilizations fail: Persian Empire (at least three imperial dynasties: 550–330 BCE, 247 BCE–224 CE, 224–651 CE); Han Dynasty (206 BCE–220 CE); Umayyad Caliphate (661–750 CE); Roman Empire (Western, 27 BCE–476 CE, and Eastern, 330–1453 CE); Mongol Empire (1206–1368 CE); Holy Roman Empire (962–1806 CE); Byzantine Empire (330–1204, 1261–1453 CE); Ottoman Empire (1299–1923 CE); Russian Empire (1721–1917 CE); and British Empire (1583–1945 CE).

    There are a number of volumes devoted to explaining why empires fail. The prescription for empire decline is arguably best addressed in Edward Gibbon's universally hailed six-volume *The History of the Decline and Fall of the Roman Empire*. The parallels to the United States are easy enough to make. Gibbons found that the Roman collapse occurred because (1) there was a decline in morals and values as divorce and violent crime rates skyrocketed, gladiatorial combat became ubiquitous, and the obsession with personal pleasure ramped up; (2) a host of environmental problems caused compromised public health, and alcoholism and disease spread by homelessness; (3) political corruption ran rampant, even to the point of selling the throne to the highest bidder in an obvious parallel to Citizens United and the influence of the Koch Brothers, Sheldon Adelson, and similar politically influential billionaires; (4) significant unemployment and a wealth and income disparity were advantageous to the wealthy agricultural class, not unlike the 1 percent/99 percent split today; (5) inflation became so rampant that bartering became common and taxing more difficult in another parallel to today's workers' real income being less now than it was in the 1970s; (6) urban decay worsened as fewer people owned homes and were forced to live in squalor-plagued apartment complexes; (7) inferior technology plagued the Romans, and their failure to invent machines to produce goods for Rome's growing population has parallels today with the United States' shipment of jobs overseas; (8) exorbitant military spending left few resources for infrastructure and public housing, not unlike the current crumbling of highways and public facilities while the United States fights in multiple wars; and (9) depletion of agricultural minerals and nutrients, as well as depleted soils and loss of top soils (metals, rare earths, fertilizers, and the like) were posing the real danger of lack of food, not to mention water.

39. "Burning Issues: Lawrence Wilkerson on U.S. Empire-Building," YouTube, April 13, 2016.
40. Jonathan Rauch, *The Constitution of Knowledge: A Defense of Truth* (Washington: Brookings Institute Press, 2021).
41. Max Fisher, *The Chaos Machine: The Inside Story of How Social Media Rewired Our Minds and Our World* (New York: Little, Brown, 2022), 7.
42. Fisher, 9, 11.
43. Martin Gilens and Benjamin I. Page, "Testing Theories of American Politics: Elites, Interest Groups, and Average Citizens," *Perspectives on Politics* 12 (September 2014): 572, 564.
44. Richard Wike, Laura Silver, Shannon Schumacher, and Aidan Connaughton, "Many in U.S., Western Europe Say Their Political System Needs Major Reform," Pew Research Center, March 31, 2021, https://www.pewresearch.org/global/2021/03/31/many-in-us-western-europe-say-their-political-system-needs-major-reform/.
45. Amitav Ghosh, *The Great Derangement: Climate Change and the Unthinkable* (Chicago: University of Chicago Press, 2016), 9.
46. Elizabeth Cripps, "Here's How to Demolish the Most Common Excuses for Climate Crisis Apathy," *Eco Investment Plans*, February 12, 2022.

Chapter Six

1. Aldo Leopold, *A Sand County Almanac: And Sketches Here and There* (New York: Oxford University Press, 1949), viii.
2. Leopold, *The Essential Aldo Leopold*, 123, 116.
3. Bill Zeedyk and Van Clothier, *Let the Water Do the Work: Induced Meandering, an Evolving Method for Restoring Incised Channels* (Santa Fe: Quivira Coalition, 2009); David L. Rosgen, *Applied River Morphology* (Pagosa Springs: Wildland Hydrology, 1996); Thomas R. Biebighauser, *Wetland Drainage, Restoration and Repair* (Lexington: University Press of Kentucky, 2007).
4. Joy B. Zedler, "The Continuing Challenge of Restoration," in *The Essential Aldo Leopold: Quotations and Commentaries*, ed. Curt Meine and Richard L. Knight (Madison: University of Wisconsin Press, 1999), 116–19, 124, 116.
5. Nina Leopold Bradley, "A Sense of Place," *Leopold Outlook* 17, no. 2 (Fall/Winter 2017): 7.
6. Bradley, 5.
7. Susanne Simard, *Finding the Mother Tree: Discovering the Wisdom of the Forest* (New York: Alfred A. Knopf, 2021); Robin Wall Kimmerer, *Braiding*

Sweetgrass: Indigenous Wisdom Scientific Knowledge and the Teachings of Plants (Minneapolis: Milkweed, 2013).
8. Zedler, "Continuing Challenge," 124.
9. Zeedyk and Clothier, *Let the Water Do the Work*.
10. James R. Kristofic, "Riparian Restoration Guru," *High Country News*, July 3, 2008.
11. Albert Einstein, BrainyQuote, https://www.brainyquote.com/quotes/albert_einstein_385842.
12. Rattan Lal, email, 2016.
13. Mike Fugali, email, April 27, 2019.
14. Barry Lopez, "An Era of Emergencies Is upon Us and We Cannot Look Away," *Literary Hub*, December 24, 2020, https://lithub.com/barry-lopez-an-era-of-emergencies-is-upon-us-and-we-cannot-look-away.

Chapter Seven

1. "Consultation: Nature and Net Zero," World Economic Forum, January 2021, 6.
2. Kristin Ohlson, *The Soil Will Save Us* (New York: Rodale, 2014); Douglas W. Tallamy, *Nature's Best Hope: A New Approach to Conservation That Starts in Your Yard* (Portland: Timber Press, 2019).
3. Montgomery, *Dirt*, 115.
4. Ohlson, *The Soil Will Save Us*, 149.
5. Noga Arikha, "Presentism," in *What Should We Be Worried About? Real Scenarios That Keep Scientists Up at Night*, ed. John Brockman (New York: Harper Perennial, 2014), 427.
6. Nicholas Humphrey, "Fast Knowledge," in *What Should We Be Worried About? Real Scenarios That Keep Scientists Up at Night*, ed. John Brockman (New York: Harper Perennial, 2014), 454–55.
7. I make two assumptions in this book that are well established but about which there is recent, reasoned disagreement. Both assumptions are challenged by current scholarship and politics. I suspect that these new disputing positions will eventually be acknowledged as correct, but there is no need to clutter this book by addressing these questions.

   Assumption 1: Despite the flaw of slavery, our Founding Fathers envisioned a democratic republic, a nation based on the idea of equality—life, liberty and the pursuit of happiness—where everyone has the opportunity to succeed, the so-called American Dream.

   Disputing position: "Minority rule is a feature, not a bug of the US constitutional system. . . . The framers of the constitution never intended that every American would have a vote, much less a vote of equal value. They created a system to regulate a society ruled by and for white male settlers engaged in the conquest and subordination of the country's Indigenous and Black populations" (Tony Karon, "Could Chile Show the United States How to Rebuild Its Democracy?" *The Guardian*, December 23, 2021).

Assumption 2: Human history developed in three phases: (1) hunter and foragers, fishers and fowlers, (2) agriculturists (the Agricultural Revolution), and (3) industrialists (the Industrial Revolution).

Disputing position: "Human societies before the advent of farming were not confined to small egalitarian bands. On the contrary, the world of hunter-gatherers as existed before the coming of agriculture was one of bold social experiments, resembling a carnival parade of bold political forms, far more than it does the drab abstraction of evolutionary theory. Agriculture, in turn, did not mean the inception of private property, nor did it mark the irrevocable step toward inequality. In fact, many of the farming communities were relatively free of ranks and hierarchies" (Graeber and Wengrow, *The Dawn of Everything*, 4).

The choice of either alternative in these two areas of disagreement does not affect the application of the theories and recommendations put forth in this book. Because it's convenient, I've opted to accept both assumptions as correct.

8. Assume one calendar year for the existence of Earth to date: on December 31 at the end of the year, the first humans appear at 11:00 p.m., and agriculture begins at two minutes before midnight.
9. Kristin Ohlson, "Where Did All the Carbon Go?" *Leopold Outlook*, Summer 2015, 4.
10. This principle is attributed to fourteenth-century English theologian William of Ockham, and while it is not always the correct solution, because photosynthesis is so fundamental to life, its application surely fits.
11. Rattan Lal, email, 2016.
12. Ohlson, *The Soil Will Save Us*, 27–28.
13. Griscom et al., "Natural Climate Solutions," 1.
14. Griscom et al.
15. Griscom et al., 1–6.
16. Joseph E. Fargione et al., "Natural Climate Solutions in the United States," *Science Advances*, November 14, 2018.
17. K. Gravuer, S. Gannet, and H. Throop, "Organic Amendment Additions to Rangelands: A Meta-Analysis of Multiple Ecosystem Outcomes," *Global Change Biology*, January 19, 2018, 1153.
18. For the first of several studies to scientifically quantify the benefits and down channel benefits of grade-control structures, see L. M. Norman, R. Lal, E. Wohl, E. Fairfax, A. C. Gellis, and M. M. Pollock, "Natural Infrastructure in Dryland Streams (NIDS) Can Establish Regenerative Wetland Sinks That Reverse Desertification and Strengthen Climate Resilience," *Science of the Total Environment* 849 (2022), https://doi.org/10.1016/j.scitotenv.2022.157738. See also N. R. Wilson and L. M. Norman, "Five Year

Analyses of Vegetation Response to Restoration Using Rock Detention Structures in Southeastern Arizona, United States," *Environmental Management* 71 (2022), https://doi.org/10.1007/s00267-022-01762-0.

19. Ohlson, *The Soil Will Save Us*, 39, 84.
20. Christine Jones, "The Back Forty Down Under: Adapting Farming to Climate Variability," *Quivira Coalition Journal* 25 (February 2010): 15.
21. Disclosure: We have an investment in a local biochar production company.
22. Courtney White and Michael Pollan, *Grass, Soil, Hope: A Journey Through Carbon Country* (White River Junction: Chelsea Green, 2014), 7.
23. Jones, "The Back Forty Down Under," 11.
24. Howie Hawkins, "350 ppm, 1°C, and 100% by 2030: The Climate Science Behind the Targets," Howie Hawkins for Governor, May 12, 2019, https://www.howiehawkins.org/350_ppm_1_c_and_100_by_2030_the_climate_science_behind_the_targets.
25. Millennium Ecosystem Assessment, *Ecosystems and Human Well-Being Synthesis* (Washington, DC: Island Press, 2005); White and Pollan, *Grass, Soil, Hope*, 22.
26. Ohlson, *The Soil Will Save Us*, 17.
27. David Gori, personal communication, 2014.
28. White and Pollan, *Grass, Soil, Hope*, xiv.
29. Sasha Harris-Lovett, "California Ranchers Tackle the Climate Crisis One Pasture at a Time," *Grist*, July 2, 2014, https://grist.org/climate-energy/california-ranchers-tackle-the-climate-crisis-one-pasture-as-a-time.
30. John L. Thomas, *A Country in the Mind: Wallace Stegner, Bernard DeVoto, History, and the American Land* (New York: Routledge, 2000), 109. The full quote: "I have been convinced for a long time that what is miscalled the middle of the road is actually the most radical and the most difficult position—much more difficult and radical than either reaction or rebellion."

Chapter Eight

1. Emmanuel Rivière, "Sharing the Responsibility for Climate Action: An Individual and Collective Commitment," *Kantar Public*, https://kantar.turtl.co/story/public-journal-04/page/3/1.
2. Rivière, 12, 9, 5.
3. Jon Henley, "Many Europeans Want Climate Action—but Less So If It Changes Their Lifestyle, Shows Poll," *The Guardian*, May 2, 2023.
4. Kayla Epstein, "Greta Thunberg Wants You to Listen to The Scientists, Not Her," *Science Alert*, September 19, 2019, https://sciencealert.com/greta-thunberg-wants-you-to-listen-to-scientists-not-her.
5. Griscom et al., "Natural Climate Solutions," 1.
6. New Mexico Interagency Climate Change Task Force, *Progress and Recommendations*, 2021.

7. Damian Carrington, "Tree Planting 'Has Mind-Blowing Potential' to Tackle Climate Crisis," *The Guardian*, July 4, 2019; Umair Irfan, "Restoring Forest May Be Our Most Powerful Weapon in Fighting Climate Change," *Vox*, July 4, 2019.
8. Tallamy, *Nature's Best Hope*, 110.
9. Uloma Uche, "Pregnant with PFAS: The Threat of 'Forever Chemicals' in Cord Blood," Environmental Working Group, September 2022.
10. Wilson, *Half-Earth*, 72.
11. "John Galt's Speech from Ayn Rand's *Atlas Shrugged*," *Amber*, www.amberandchaos.net/?pageid=73.
12. Charles Darwin, *The Descent of Man: Selection in Relation to Sex* (New York: D. Appleton, 1896), 132.
13. Matt Ridley, *The Origins of Virtue: Human Instincts and the Evolution of Cooperation* (New York: Viking, 1997), 146.
14. Edward O. Wilson, *The Social Conquest of the Earth* (New York: Liveright, 2012), 174, 162–63; Rutger Bregman, *Humankind: A Hopeful History* (New York: Little, Brown, 2019). For more information on this topic, see Stanley Milgram's Yale University Shock Machine Experiment, Philip Zimbrado's Stanford Prison Experiment, Jared Diamond's scholorship on Easter Island, Dawkin's selfish gene theory, and Yuval Noah Harari's understanding of human nature as set out in *Sapiens: A Brief History of Humankind* (New York: HarperCollins, 2015). Even the "bystander effect" is based on flawed studies and seriously erroneous thinking about human nature. Instead, Bregman proposes just the opposite: "survival of the friendliest" (*Humankind*, 64).
15. "Global Destruction of Nature Being Subsidized by $1.8 Trillion Annually," *Business for Nature*, February 17, 2022.
16. Brian Kahn, "The World Blows over $5 Trillion a Year on Oil and Gas Subsidies: Report," *Gizmodo*, May 19, 2019.
17. Joel Jaeger, Ginette Walls, Ella Clarke, Juan-Carlos Altamirano, Ayra Harsono, Helen Mountford, Sharan Burrow, Samantha Smith, and Alison Tate, "The Green Jobs Advantage: How Climate-Friendly Investments Are Better Job Creators," World Resources Institute, October 18, 2021.
18. Bill McKibben, "Global Warming's Terrifying New Math," *Rolling Stone*, July 19, 2012.
19. Quoted in Robert P. Murphy, "The Economics of Climate Change," *Econlib*, July 6, 2009.
20. Naomi Klein, *This Changes Everything: Capitalism vs. the Climate* (New York: Simon and Schuster, 2014), 102.
21. Lou Helmuth, deputy director, Our Children's Trust, email, January 18, 2022.
22. *Juliana v. The United States of America*, 6:15-cv-1517-TC, Order and Findings & Recommendation, April 8, 2016, 10, 1.

23. *Juliana v. The United States of America*, 1.
24. Amy Goodman, "Landmark Climate Lawsuit: Meet the Youth Activists Suing the U.S. Government and the Fossil Fuel Industry," *Democracy Now!*, interview of Julia Olson, April 14, 2016.
25. McKibben, Klein, and Gregory quoted in "Victory in Landmark Climate Case!" *Our Children's Trust*, April 8, 2016.
26. *Democracy Now!*, April 14, 2016. Emphasis added.
27. *Juliana v. The United States of America*, Opinion and Order, November 10, 2016, 32.
28. *Held v. State of Montana*, Cause No. CDV-2020–308, 13.
29. *Sharma by Her Litigation Representative Sister Marie Brigid Arthur v. Minister for the Environment*, FCA 560 (2021), 90.
30. *Cherniak v. Kate Brown*, Brief of Amici Curiae, Supreme Court No. S066564, March 22, 2019, 3.
31. For information on this issue, see *The Wave*, a newsletter on climate litigation.
32. Marco Grasso and Richard Heede, "Time to Pay the Piper: Fossil Fuel Companies' Reparations for Climate Damages," *One Earth*, May 18, 2023.
    The report summarizes several of the essentials of this book:

> While the Global North's historical carbon emissions have exceeded their fair share of the planetary boundary by an estimated 92%, the impacts of climate breakdown fall disproportionally on the Global South, which is responsible for a trivial share—Africa, Asia, and Latin America contribute only 8%—of excess emissions. At the same time, the world's richest 1% of the population contributed 15% of emissions between 1990 and 2015, more than twice as much as the poorest 50%, who contributed just 7% but who suffer the brunt of climate harm. This inequity is exacerbated by poorer societies' lack of resources to adapt to climate impacts and by the persistent reluctance of the Global North to provide them with the necessary funding and assistance as required by the principle of common but differentiated responsibilities and respective capabilities.

33. Center for International Environmental Law, https://www.ciel.org.

Chapter Nine

1. Courtney White, "An Unprecedented Future," in *Conservation for a New Generation: Redefining Natural Resources Management*, ed. Richard Knight and Courtney White (Washington, D.C.: Island Press, 2009), 286.
2. Rauch, *The Constitution of Knowledge*, 41.

3. E. O. Wilson, Quoteish, 20 Restoration Quotes, https://quoteish.org/2019/12/restoration-quotes.html.
4. Stegner, *Where the Bluebird Sings*, 117.
5. Peninah Neimark and Peter Rhoades Mott, eds., *The Environmental Debate: A Documentary History* (Amenia, NY: Grey House, 2010), 94.
6. Joseph Edward de Steiguer, *The Origins of Modern Environmental Thought* (Tucson: University of Arizona Press, 2006), 59.
7. Bob Weidner, "Doing Well Is Doing Good," *Metals Service Center Institute*, August 13, 2019.
8. Genesis 1:28.
9. Gifford Pinchot, *The Training of a Forester* (Philadelphia: J. B. Lippincott, 1914), 13.
10. Stegner, *Where the Bluebird Sings*, 131.
11. Stegner, *The Sound of Mountain Water*, 145–54, 146.
12. Terry Tempest Williams, Goodreads, https://www.goodreads.com.
13. Marnie Walker Gaede, ed., *Images from the Great West: Photographs by Marc Gaede* (La Cañada: Chaco Press, 1990), 8.
14. Leopold, *A Sand County Almanac*, 204.
15. J. Baird Callicott, "The Land Aesthetic," in *Companion to* A Sand County Almanac*: Interpretive and Critical Essays*, ed. J. Baird Callicott (Madison: University of Wisconsin Press, 1987), 157.
16. William R. Jordan III, *The Sunflower Forest: Ecological Restoration and the New Communion with Nature* (Berkeley: University of California Press, 2003), 30.
17. Leopold, *River of the Mother of God*, 185.
18. Leopold, 223.
19. Leopold, 186, 191. Emphasis in original.
20. Jordan, *The Sunflower Forest*, 30.
21. Ronald W. Clark, *The Survival of Charles Darwin: A Biography of a Man and an Idea* (New York: Random House, 1984), 145.
22. Jack Miles, *Religion as We Know It: An Origin Story* (New York: W.W. Norton, 2019).
23. Anna L. Peterson, *Being Human: Ethics, Environment, and Our Place in the World* (Berkeley: University of California Press, 2001), 30, 34.
24. Jeremy Lent, *The Patterning Instinct: Cultural History of Humanity's Search for Meaning* (Amherst: Prometheus, 2017), 121.
25. Peterson, *Being Human*, 47.
26. Lent, *Patterning Instinct*, 123.
27. Berry, *The World-Ending Fire*, 27.
28. Jordan, *The Sunflower Forest*, 37.
29. George Monbiot, *Out of the Wreckage: A New Politics for an Age of Crisis* (London: Verso, 2017), 14.
30. Chris Hedges, *Wages of Rebellion: The Moral Imperative of Revolt* (New York: Nation, 2015), 41; Curt Meine, "Land, Ethics, Justice, and Aldo Leopold," *Socio-Ecological Practice Research* 4 (2022): 167.

31. Hanna Krall, *Shielding the Flame: An Intimate Conversation with Dr. Marek Edelman, the Last Surviving Leader of the Warsaw Ghetto Uprising*, trans. Joanna Stasinska and Lawrence Weschler (New York: Henry Holt, 1986), 37.
32. Wallace Stegner, "The Legacy of Aldo Leopold," in *Companion to* A Sand County Almanac*: Interpretive and Critical Essays*, ed. J. Baird Callicott (Madison: University of Wisconsin Press, 1987), 233; Wallace Stegner, *Marking the Sparrow's Fall: Wallace Stegner's American West*, ed. Page Stegner (New York: Henry Holt, 1998), 174.
33. Thích Nhất Hạnh, *Love Letter to the Earth* (Berkeley: Parallax, 2013), 57, 83, 10, 8, 9.
34. Ghosh, *The Great Derangement*, 132.
35. IPCC, "Summary for Policymakers of IPCC Special Report on Global Warming of 1.5°C Approved by Governments," October 8, 2018, https://www.ipcc.ch/2018/10/08/summary-for-policymakers-of-ipcc-special-report-on-global-warming-of-1-5c-approved-by-governments/.
36. "WHO Coronavirus (COVID-19) Dashboard," World Health Organization, May 17, 2023; "Estimating Excess Mortality due to the COVID-19 Pandemic: A Systematic Analysis of COVID-19-Related Mortality," *Lancet*, April 16, 2022.
37. Camilo Mora, Tristan McKenzie, Isabella M. Gaw, Jacqueline M. Dean, Hannah von Hammerstein, Tabatha A. Knudson, Renee O. Setter, Charlotte Z. Smith, Kira M. Webster, Jonathan A. Patz, and Erik C. Franklin, "Over Half of Known Human Pathogenic Diseases Can Be Aggravated by Climate Change," *Nature Climate Change* 12 (2022): 869–75.
38. Stegner, *Marking the Sparrow's Fall*, 179.
39. Stegner, 181.
40. Stegner, 175 and 179.
41. Graeber and Wengrow, *The Dawn of Everything*, 63.
42. Chiara Bottici, "Democracy and the Spectacle: On Rousseau's Homeopathic Strategy," *Philosophy and Social Criticism* 41, no. 3 (2015): 237, 238.
43. Guy Debord, *The Society of the Spectacle* (New York: Zone, 1996), 16.
44. Bottici, "Democracy and the Spectacle," 246.
45. Nick Estes, *Our History Is the Future: Standing Rock Versus the Dakota Access Pipeline and the Long Tradition of Indigenous Resistance* (London: Verso, 2012), 260.
46. This three-part expression originates from various sources of advice on manners, sermons, and suitable subjects of civilized conversation, personal character, and conduct. It originally spoke in terms of "ideas, things and people." In a 1901 autobiography by Charles Stewart, he recounted a conversation of dinner guests he overheard as a child in London where history scholar Henry Thomas Buckle—whose conversation, Stewart maintained, was always on a high level—said that "men and women range themselves into three classes or orders of intelligence; you can tell the lowest class by their habit of always talking about persons; the next by the fact that their habit is always to converse about things; the highest

by their preference for the discussion of ideas." Over time, "events" was substituted for "things," and in a 1987 Kansas newspaper, the saying was attributed to Eleanor Roosevelt, with whom attribution now remains. For further details, see "Great Minds Discuss Ideas; Average Minds Discuss Events; Small Minds Discuss People," Quote Investigator, https://quoteinvestigator.com//2014/11/18/great-minds/.

47. Peterson, *Being Human*, 75.
48. Leopold, *River of the Mother of God*, 185. Emphasis added.
49. J. Baird Callicott, "The Conceptual Foundations of the Land Ethic," in *Companion to* A Sand County Almanac*: Interpretive and Critical Essays*, ed. J. Baird Callicott (Madison: University of Wisconsin Press, 1987), 195.
50. Leopold, *A Sand County Almanac*, 224–25.
51. Callicott, "Conceptual Foundations."
52. Aldo Leopold, "Some Fundamentals of Conservation in the Southwest," *Environmental Ethics* 1 (1979): 139–40. Emphasis added.
53. Callicott, "Conceptual Foundations," 201.
54. Stegner, *Marking the Sparrow's Fall*, 176.

Chapter Ten

1. Quoted in Michelle Hall Kells, *Vicente Ximenes, LBJ's Great Society, and Mexican American Civil Rights Rhetoric* (Carbondale: Southern Illinois University Press, 2018), 88.
2. Jack Cargill, "Empire and Opposition: The Salt of the Earth Strike," in *Labor in New Mexico: Unions, Strikes, and Social History since 1881*, ed. Robert Kern (Albuquerque: University of New Mexico Press, 1983), 183.
3. Ellen R. Baker, *On Strike and on Film: Mexican American Families and Blacklisted Filmmakers in Cold War America* (Chapel Hill: University of North Carolina Press, 2007), 104.
4. *Jencks v. United States*, 353 U.S. 657, 1957. Almost as important as the United States Supreme Court's 1966 *Miranda* right to remain silent and legal counsel decision and the 1963 *Brady v. Maryland* right to exculpatory evidence cases, the *Jencks* decision on the Sixth Amendment right to confront and cross-examine witnesses is at the very core of the Fourteenth Amendment's protections against prosecutorial and governmental overreach. These three decisions—the rights to legal counsel, to remain silent, to have access to exculpatory evidence and to cross-examine witnesses—form the core of United States criminal jurisprudence.
5. Baker, *On Strike and on Film*, 4.
6. Cargill, "Empire and Opposition"; James J. Lorence, *The Suppression of* Salt of the Earth*: How Hollywood, Big Labor, and Politicians Blacklisted a Movie in Cold War America* (Albuquerque: University of New Mexico Press, 1999); James J. Lorence, *Palomino, Clinton Jencks, and Mexican-American*

Unionism in the American Southwest (Urbana: University of Illinois Press, 2013); Raymond Caballero, *McCarthyism vs. Clinton Jencks* (Norman: University of Oklahoma Press, 2019).

7. Baker, *On Strike and on Film*, 4.
8. Cargill, "Empire and Opposition," 253.
9. Cargill, 194–95.
10. Lorence, *Palomino*, 23, xvii.
11. Soviet Premier Nikita Khrushchev's "We will bury you" remark was seen by most Americans as a nuclear threat, but a closer examination suggests that the remark may well have been little more than a bungled effort at rhetorical flair and far less threatening than commonly understood in the West. What he actually said while addressing Western ambassadors during a 1956 reception at the Polish embassy in Moscow was "Whether you like it or not, history is on our side. We will dig you in." Khrushchev was rephrasing the popular Marxist saying: "The proletariat is the undertaker of capitalism." This apothegm shadows the assertion in the *Communist Manifesto* that the bourgeoisie will produce its own "gravediggers," the working class will inevitably bury those of the dominating capitalist upper class, and capitalist economies will inevitably collapse of their own accord.
12. Madam Millie, a.k.a. Mildred Clark Cusey, was an orphan with a tuberculosis-afflicted sister whom she cared for until her sister's death. Millie came west to waitress as a Harvey Girl on the Aitchison, Topeka, and Santa Fe Railroad and became a highly successful and notorious businesswoman. According to Max Evans, "During the vicious copper strike of the [early] 1950s, Hispano strikers had very little money saved, and they ran out of food. Millie fed their children for months. It was only white bread, bologna, and cheese, but it was food. . . . Despite possible boycotts from her wealthy patrons, many of whom were copper company middle managers, Millie had her own set of loyalties, her own values. She fed the poor. She fed the families of the strikers" (Max Evans, *Madam Millie* [University of New Mexico Press, Albuquerque, 2002], xvii).
13. *Bringing* Salt of the Earth *Home: A 50th Anniversary Symposium* (Silver City: Silver City Museum Historical Society, 2004), film, 1:29:40.
14. *Bringing* Salt of the Earth *Home*, quoting Jencks, 1:33:15.
15. Lorence, *Suppression*, 30. Emphasis added.
16. *Bringing* Salt of the Earth *Home*, quoting Jencks, 1:34:40.
17. *Bringing* Salt of the Earth *Home*, quoting Jencks, 1: 35:12.

The process of "consciousness raising" had surely occurred throughout the world well before the 1960s. However, the process had not been crystallized, and the term itself was not coined until late 1967, by the New York Radical Women and other groups, and thereafter the process quickly spread throughout the United States. According to Susan Brownmiller, "In the Old Left, they used to say that workers don't know they're oppressed, so we have to raise their consciousness. One night at a meeting I said, 'Would everybody please give me an example from their own life on how they experienced oppression as a woman? I need to hear it to raise my own consciousness.' Kathy was sitting behind me and the words rang in her mind. From then on she sort of made it an institution and called it consciousness-raising" (Susan Brownmiller, *In Our Time: Memoir of a Revolution* [New York: Dial, 1999]).

On Thanksgiving 1968, Kathie Sarachild presented A Program for Feminist Consciousness Raising at the First National Women's Liberation Conference near Chicago, Illinois, in which she explained its principles and outlined a program for raising consciousness. From there it morphed into a central component of the Feminist Movement (Kathie Sarachild, "Consciousness-Raising: A Radical Weapon," Redstockings, 1973, https://womenwhatistobedone.files.wordpress.com/2013/09/1973-consciousness-raising-radical-weapon-k-sarachild-redstockings.pdf).

Of course, the women's movement has since evolved yet again and exploded with the #MeToo movement, which that spread virally in October 2017 and was followed by a number of public revelations of sexual assault and harassment of women by men (and vice versa) in positions of power, a movement that has cost the careers of a number of high-profile men and surely has many others waiting for the hammer to drop—quite a distance from Teresa Vidal's suggestion.

18. *Bringing* Salt of the Earth *Home*, quoting Jencks, 1:32:46.
19. Lorence, *Suppression*, 33. Emphasis added.
20. Cargill, "Empire and Opposition," 214.
21. Leslie Goforth played a curious part in the Empire-*Salt* saga. On the one hand, in the context of their hostility toward Bartley McDonald, the Mexican community catapulted Goforth into the office of Grant County sheriff, winning the election by the thinnest of margins after several recounts. But Goforth sided with the Empire Zinc Company.

Local 890 members should have known they were spinning their wheels and trading for a "pig in a poke" in swapping Goforth for McDonald. Even before the recount was official, Goforth made it clear that he was "under no obligation to the International Mine, Mill & Smelter Workers, Local 890 for their unsolicited support during the campaign" (*Silver City Enterprise*, November 16, 1950). Despite union support that clearly led to a change in sheriffs, Goforth promptly made it clear that there would be no change in policy. Just two years earlier, McDonald had bested Goforth by 663 votes with 56 percent of the vote (3,132 to 2,469). But in this election, even though "in all other county offices the pendulum swung to the democratic side of the ballot sending candidates into office with overwhelming majorities . . . in probably the closest election contest ever held in Grant

County," Goforth squeaked by with an eight-vote majority (2,979 to 2,971), or 50.067 percent of the vote (*Silver City Enterprise*, November 9, 1950).

Also on November 9, 1950, the *Silver City Daily Press* reported that Goforth was up by only two votes. That number was later upped to three votes. Both candidates sought a recount, and it was rumored that eight votes for the incumbent were discarded as illegal because of unusual markings. It was next reported that Goforth was a "scant three votes ahead of the incumbent," which was followed by a recount of the Santa Rita district insisted upon by McDonald, where he "lost 10 of his original 77" votes and was finally forced to give up his Office by eight votes (*Silver City Enterprise*, November 30, 1950). Had the Santa Rita votes not been recounted, McDonald may well have won by an even slimmer two-vote margin.

In terms of what locals thought of Leslie Goforth, when the time came to film *Salt of the Earth*, the director was forced to hire a professional actor because no one locally was willing to take the part of Goforth.

22. Howard Zinn was the History Department chair of Spelman College in 1963 when he was fired because of his civil rights activities. In 2005 Zinn—"deeply honored to be invited back after forty-two years"—gave a commencement address in which he encouraged graduating seniors not to be "discouraged by the way the world looks at this moment," to become and remain engaged. Published in *Original Zinn: Conversations on History and Politics*, the quote below the Bayard mural is typical. Zinn, capturing the spirit of the ordinary women and men who took to the street for so many months, took huge risks and refused to give up. Howard Zinn, "Against Discouragement," https://www.howardzinn.org/collection/against-discouragement/.
23. Lorence, *Palomino*, 97, 42.
24. Lorence, 7, 4, 31.
25. For those who grew up watching Efrem Zimbalist Jr. as the go-to agent of *The FBI* television series that ran from 1965 to 1974, Torrez's claim may not ring true. Despite the once favorable reputation of J. Edgar Hoover and his FBI, the truth is that Hoover had long been operating his own secret shadow Bureau of Investigation outside the law. Hoover's secret and illegal counterintelligence program was fully operational by the time of the Empire mine strike—a "dirty tricks" operation eventually operating under the name of COINTELPRO—established as early as 1956 to investigate and disrupt dissident political groups in the United States. COINTELPRO was first used to disrupt the Communist Party as Hoover purposely exaggerated the threat of communism to ensure public and financial support for the FBI. Hoover's tragic misuse of such a powerful institution is thoroughly chronicled in Betty Medsger's *The Burglary: The Discovery of J. Edgar Hoover's Secret FBI* (New York: Vintage, 2014), in which she recounts the remarkable tale of how in 1971, eight ordinary cit-

izens, calling themselves the Citizens Commission to Investigate the FBI, burglarized the FBI building in Media, Pennsylvania, and then disclosed untold abuses to the media. They were never apprehended.

26. Lorence, *Palomino*, xiv.
27. Lorence, 201.

Chapter Eleven

1. Orwell, *Orwell on Truth*, 123.
2. G. S. Callendar, "The Artificial Production of Carbon Dioxide and Its Influence on Temperature," *Quarterly Journal of the Royal Meteorological Society*, 1938, 223.
3. Callendar.
4. Spencer R. Weart, "The Discovery of Global Warming: Bibliography by Year Through 2001," Center for History of Physics of the American Institute of Physics, August 2021, https://history.aip.org/climate/iindex.htm. Weart has a book by the same name: Spencer R. Weart, *The Discovery of Global Warming: Revised and Expanded Edition* (Cambridge: Harvard University Press, 2008).
5. Edwin S. Grosvenor and Morgan Wesson, *Alexander Graham Bell: The Life and Times of the Man Who Invented the Telephone* (New York: Harry N. Abrams, 1997), 274.
6. This ruse was a key tactic disclosed in Naomi Oreskes and Erik M. Conway's *Merchants of Doubt* (New York: Bloomsbury, 2010), 10.
7. 14th International Conference on Climate Change, "The Great Reset: Climate Realism vs. Climate Socialism," https://stayhappening.com/e/14th-international-conference-on-climate-change-E2lSTRYD1KS. According to H. Sterling Burnett, "The program responds to international leaders' push to exploit the supposed existential threat of climate change to get people to 'reimagine capitalism,' in the words of Klaus Schwab, founder and executive chairman of the World Economic Forum. As described in numerous documents and discussions, the aim of those pushing the Great Reset (GR) is to impose socialism worldwide, especially on energy production and use, which is the foundation of the modern world" ("ICCC-14 Warns of Global Elitists' Great Reset," Heartland Institute, September 16, 2021).
8. Julia Olson and Philip Gregory, "Introduction," in James Gustave Speth, *They Knew: The U.S. Federal Government's Fifty-Year Role in Causing the Climate Crisis* (Cambridge: MIT Press, 2021), ix.
9. Union of Concerned Scientists, "The Climate Deception Dossiers," 2005, www.ucsusa.org/DecadesofDesception.
10. Quoted in Eoin Higgins, "Coal Knew Too: Explosive Report Shows Industry Was Aware of Climate Threat as Far Back as 1966," *Common Dreams*, November 22, 2019.

11. Abrahm Lustgarden, "The Great Climate Migration," *New York Times*, July 7, 2023.
12. "Smoke and Fumes: The Legal and Evidentiary Basis for Holding Big Oil Accountable," Center for International Environmental Law, November 16, 2017, https://www.ciel.org/news/smoke-and-fumes-2/.
13. "Smoke and Fumes."
14. Victor Sebestyen, "Bannon Says He's a Leninist: That Could Explain the White House's New Tactics," *The Guardian*, February 6, 2017.
15. Monique Beals, "Bannon Says He Discussed How to 'Kill this Administration in the Crib' with Trump Before Jan. 6," *The Hill*, September 23, 2021.
16. Evelyn Blackwell, "Steve Bannon Calls For 'Shock Troops' to 'Deconstruct' State as GOP Takes Oval Office," *World News Era*, October 2, 2021.
17. Graeme Massie, "Steve Bannon Pledges 20k 'Shock Troops' Ready to Go as He Rants That 'We Control the Country,'" *The Independent*, reviewed October 5, 2021.
18. "Scared Scientists Fear Global Warming," Gogglepix, All Things Art & Photography, September 10, 2014. This article is a "great example of how the spoken word and a photograph can still make a difference in this world. By making it personal, photographer Nick Bowers along with Celine Faledam and Rachel Guest interviewed and photographed a group of scientists about climate change. If the conversations don't scare the hell out of you nothing will."

Chapter Twelve

1. Susan Neiman, *Learning from the Germans: Race and the Memory of Evil* (New York: Farrar, Straus and Giroux, 2019), 65.
2. "Richest 1% Bag Nearly Twice as Much Wealth as the Rest of the World Put Together over the Past Two Years," Oxfam International, January 16, 2023.
3. Chris Hedges, *RUMBLE with Michael Moore*, September 2, 2021, https://rumble.media/episode209transcript/.
4. Marya and Patel, *Inflamed*, 64.
5. Cripps, "Here's How to Demolish the Most Common Excuses."
6. This quote is commonly attributed to Dante but apparently evolved via "a multistep process from a changing and imperfect interpretation of Dante's work. In 1915 Theodore Roosevelt accurately wrote that Dante had 'reserved a special place of infamy' for neutral angels. In 1917 a religious orator named W. M. Vines incorrectly stated that Dante had placed neutral individuals 'in the lowest place in hell.' In 1944 the spiritual writer Henry Powell Spring penned a book of aphorisms that included a statement ascribed to Dante that closely matched the modern quotation. John F. Kennedy used the saying several times in speeches in the 1950s and

later. Kennedy also attributed the remark to Dante" (Quote Investigator, https://quoteinvestigator.com/2015/01/14/hottest/).

7. Mario Vargas Llosa, *Making Waves* (New York: Farrar, Straus, and Giroux, 1996), 297.
8. *State of Washington v. Kenneth A. Ward*, Court of Appeals of the State of Washington, April 8, 2019, 5.
9. William M. Adler, *The Man Who Never Died: The Life, Times, and Legacy of Joe Hill, American Labor Icon* (New York: Bloomsbury, 2011), 186.
10. Emily Johnston, "I Shut Down an Oil Pipeline—Because Climate Change Is a Ticking Bomb," *The Guardian*, November 24, 2017. Emphasis in original.
11. Erin Grady, "Victory for Valve Turners in Minnesota!" Civil Liberties Defense Center, October 9, 2018,
12. Grady.
13. Berry, *The World-Ending Fire*, 67. Emphasis added.
14. Steffen quoted in Stephen Bau, *Builders Collective* (blog), November 8, 2019, https://bauhouse.medium.com/you-never-change-things-by-fighting-the-existing-reality-3e34a55ab653.
15. Matthew Taylor, "Global Climate Strike: Thousands Join Coordinated Action Across the World," *The Guardian*, September 24, 2021.
16. Ieshia Evans, "I Wasn't Afraid. I Took a Stand in Baton Rouge Because Enough Is Enough," *The Guardian*, July 22, 2016.
17. Jane Doe, personal conversation with the author, September 11, 2011.
18. "World Energy Outlook Report," International Energy Agency, June 2020.
19. "World Energy Outlook Special Report," International Energy Agency, June 2021.
20. Goodreads, https://www.goodreads.com/quotes/13119-you-never-change-things-by-fighting-the-existing-reality-to.
21. Miya King-Flaherty, "Sandoval People Power," *Rio Grande Sierran*, January/February/March 2018.
22. The assault on the Gila River has been relentless yet demonstrates the power of an active citizenry. In the 1970s, Grant County residents blocked the Hooker Dam proposal. Then in the 1980s, the community stopped the Conner Dam. This sordid history continued into the early 1990s with another version of the same ruse, the so-called Mangus Diversion Alternative. It too was turned back. This most recent effort to sabotage the Gila is the fourth in the last half century.

    When we lived in Casa Grande, Arizona, aggressive community opposition prevented three of three dangerous corporate intrusions into the city: the expansion of Reliant Energy's power plant in the city limits was blocked; the relocation of a polluting waterbed frame manufacturer that had been forced out of California was stopped; and an effort by American Smelting and Refining Company's (ASARCO) to convert its abandoned open-pit mine into the world's largest garbage dump was also rejected.

Each success was powered by community voice that forced the city council to reverse earlier encouragement or approval.

23. Maria Popova, "Mapping Meaning in a Digital Age," *On Being*, March 18, 2016.
24. Barry Lopez, "An Era of Emergencies Is Upon Us and We Cannot Look Away," *Literary Hub*, December 24, 2020, https://lithub.com/barry-lopez-an-era-of-emergencies-is-upon-us-and-we-cannot-look-away.
25. Rebecca Solnit, "'Hope Is an Embrace of the Unknown': Rebecca Solnit on Living in Dark Times," quoting Patrisse Cullors, *The Guardian*, July 15, 2016, https://www.guardian.com/books/2016/jul/15/rebecca-solnit-hope-in-the-dark-new-essay-embrace-unknown.
26. Lopez.
27. Albert Einstein, https://underthebluedoor.org, July 11, 2013,
28. Donna Stevens, "There Oughter Be Otters," *Carapace* 22, no, 2 (Winter 2019): 10.

Conclusion

1. Barbara Kingsolver, *Unsheltered* (New York: HarperCollins, 2018), 408–9. Emphasis in original.
2. Rabbi Eliyahu Dessler, "The Bechira Point," Tikkun Middot Project Curriculum, October 2013, https://images.shulcloud.com/428/uploads/PDFs/the-bechirah-point-rabbi-e-dessler.pdf.
3. Quoted in William Miller, "Death of a Genius: His Fourth Dimension, Time, Overtakes Einstein," *Life Magazine*, May 2, 1955.

APPENDIX A
Ciénaga Spelling, Punctuation, and Pronunciation

The Real Academia Española's *Dictionary of the Spanish Language* has this to say about the spelling of the Southwest's rare desert wetland: the term *ciénega* is derived from Latin *caenĭca, de caenum,* and *cieno.* One theory for the definition of *ciénaga* is that it comes from the expression *cién aguas,* meaning "a hundred fountains," "a hundred springs," or "one hundred waters," but linguistically, despite the term *cien* meaning "one hundred," in this context the term has nothing to do with "hundred."

Although the origin of *ciénega* and its variant *ciénaga* is not a simple one, the root is "silt," which is the meaning of *cieno,* according to the Real Academia Española. The origin of what can only be considered a colloquial definition—"one hundred waters"—is unknown to me, but that phrase is a sensible explanation for both the definition and the spelling *ciénaga* with an *a,* currently the less common spelling in the American Southwest. Certain enclaves in the Spanish-speaking world (e.g., Colombia) utilize the *a* spelling in the formal name of many swamps and bogs, and the second *e* spelling is rare. The 1958 *Cassell's Spanish Dictionary* spells *ciénaga* with the accent above the *e,* and the *a* rather than a second *e.*

Robert Julyan in *The Place Names of New Mexico* (1996) notes that although the *e* spelling had been earlier criticized, many early Spanish explorers and newcomers came from Estremadura, Spain, where *ciénega* was properly spelled with a second *e*, despite the *a* spelling being correct. In *New Mexico Place Names: A Geographical Dictionary* (1965), T. M. Pearce lists fourteen New Mexico examples of *ciénaga* usage—land grants, towns, and water features—and they are all spelled with an *a* and no acute accent above the first *e*. Pearce's book preceded Julyan's by twenty-five years and uniformly uses *a* spelling, making no mention of spelling *cienega* with the second *e*.

When looking closely at these books, it is apparent that Julyan used numerous examples from Pearce and substituted the second *e* for *a* without explanation. Two examples are Pearce's entries for ciénaga (Otero County) and ciénaga (Catron County) on page 35 of his earlier book, from which Julyan later draws, and substitutes the *e* for the *a*, on page 84 of his book. Pearce spells the *ciénagas* in Otero and Catron Counties with the *a*, yet Julyan, again, without comment, substitutes an *e* for the *a*. It is also unclear when and why the *e* spelling, represented by Julyan as "general," became so in the American Southwest.

A possible explanation for the more common *e* spelling today is that when Dean A. Hendrickson and W. L. Minckley, authors of the abstract "Cienegas: Vanishing Climax Communities of the American Southwest" (1985), first suffused the term *ciénaga* with biological significance unique to the groundwater-fed arid-land ciénagas of the American Southwest, they chose the *e* spelling that has persisted in scientific literature. I've asked Dean why they chose the second *e* spelling, and he didn't recall. Aldo Leopold used the *e* spelling and acute accent over the first *e* in his seminal 1949 *A Sand County Almanac: And Sketches Here and There*.

The spelling of *cienega* using the second *e* has become common in the United States in scientific if not popular usage and is common on many contemporary maps. The Burro Ciénaga on the Pitchfork Ranch has always been spelled with an *a*; Julyan's changes from *a* to *e* on Pearce's earlier list of ciénagas strikes me as arbitrary. *Cassell's* spells it with an *a*, thus I spell *ciénaga* with an *a* and suggest it to be the best spelling. The acute accent over the *e* is proper, although often omitted.

Similarly, pronunciation is an egg difficult to unscramble, corruption of the term having many variants: *sienaga*, *sinigie*, *senigie*, *seneca*, *sienakey*, and more. Lucinda and I can always tell when we meet someone who was raised in this area. When they learn we live on the Pitchfork Ranch, invariably, they say: "Oh, you live down by the sienakey."

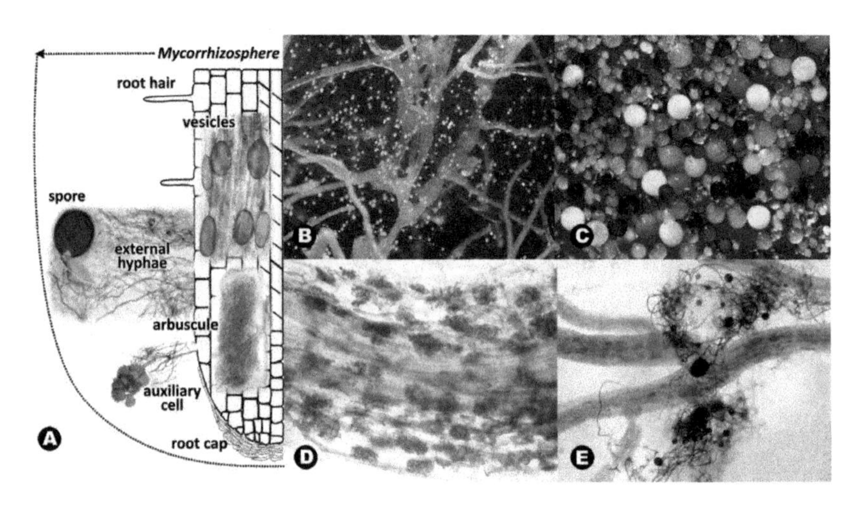

FIGURE 44 Image by Joseph Morton after the *McGraw-Hill Yearbook of Science and Technology*, 2011.

APPENDIX B
Mycorrhizal Symbiosis

ALTHOUGH THIS DETAIL IS arguably beyond the scope of this book, for the seriously curious, the following explanation scratches the surface of mycorrhizal symbiosis. It's encouraging to realize how much is now known about the subsurface world and its importance for carbon sequestration. My cousin Joe Morton, retired professor of environmental microbiology at West Virginia University, provided me with these materials that explain mycorrhizal symbiosis. Figure 44 is a collage showing the overall mycorrhizal structure and its contribution to the "mycorrhizosphere" with appropriate parts labeled. He created this image for an article in a *McGraw-Hill Encyclopedia* yearbook several years ago. Morton modified it here and provided the following description:

> Illustration of the various structures produced by soil-borne fungi that form an obligate mutualistic symbiosis with about 80 percent of all plant species. They are called "arbuscular endomycorrhizal fungi" because much of the fungus is inside the root and nutrient exchange between plant and fungus occurs at the arbuscule. Fine external hyphae explore the soil to absorb immobile phosphorus and translocate this

nutrient back to the root. This hyphal network makes up the "mycorrhizosphere" which extends far beyond the "rhizosphere" that consists of roots only. Asexual spores are produced in the soil to spread the fungus far beyond the parent plant host.

The sections in figure 44 are as follows:

- Section A: Mycorrhizal carrot roots producing abundant spores and an external hyphalnetwork of *Rhizophagus clarus* in a sterile agar medium. The roots are transformed genetically so they are immortal (growing indefinitely).
- Section B: Spores extracted from the root system of a host plant. Eight species are present: *Gigaspora gigantea* (large yellow-green), *Dentiscutata erythropa* (large dark red), *Dentiscutata heterogama* (small dark red), *Racocetra verrucosa* (large orange-brown), *Funneliformis mosseae* (smaller pale yellow brown), *Claroideoglomus etunicatum* (small orange-brown), *Rhizophagus intraradices* (small pale yellow brown) and *Acaulospora morrowiae* (small pale yellow brown also). Plants in any environment support anywhere from three to twelve species in a root system.
- Section C: Arbuscules interconnected by internal hyphae forming typical mycorrhizae in a corn roots.
- Section D: Corn mycorrhizal roots with external hyphae from which spores are forming. This species, *Septoglomus deserticola*, is common in some semiarid habitats but also is widespread globally.

Morton and a colleague submitted a paper, for a proceedings book (unpublished as of this writing) to accompany a presentation Morton gave at the Fifth International Date Palm Conference in Abu Dhabi, United Arab Emirates, in March 2014 ("Benefits of the Arbuscular Mycorrhizal Symbiosis on Productivity and Sustainability of Plants in Natural and Agricultural Communities"). In it he explains mycorrhizal symbiosis further:

The arbuscular mycorrhizal association is an essential symbiosis between a unique group of soil fungi and a majority of plant species worldwide. . . . [It] is a critical component to the survival and longevity of a vast number of plant species, including all crop plants. . . . Arbuscular mycorrhizal fungi evolved four hundred million years ago, at about the same time plants appeared on land. Since those early land plants did not have a true root system for absorption of nutrients, the presence of Arbuscular mycorrhizal fungi is hypothesized to have been crucial to evolution of plants on land. A survey of the plant kingdom within the framework of their evolution suggests that all of the earliest land plants formed an arbuscular mycorrhizal association . . . and that over the intervening eons some plants either evolved different kinds of mycorrhizal associations . . . or they lost the symbiosis completely. Arbuscular mycorrhizal fungi can obtain carbon for growth and reproduction only through specialized structures called "arbuscules" that form at a close interface between plant and fungus in root cortical cells. As a result, these fungi cannot grow apart from a host plant. Each arbuscule appears to fill a cell but it resides between the wall and membrane to maintain functionality. Phosphorus tightly bound to soil particles is absorbed by an extensive network of finely branched hyphae that fills the rhizosphere and it is from these hyphae that asexual spores are formed for dispersal.

APPENDIX C
Personal Climate Pledge

1. Completely eliminate bacon and eat meat only once a week; eat vegetarian as often as possible.
2. Mend clothing and take all used items to a secondhand store.
3. Recycle everything possible.
4. Repair whenever possible and avoid throwing away.
5. Telephone and send follow-up letters to companies to explain how they can make their packaging environmentally friendly or encourage recycling.
6. Plan additional time to travel by train rather than airplane. Limit air travel to one short-haul flight every two to three years and one long-haul flight every eight years.
7. Limit the purchase of new clothes and shoes. Donate infrequently used items, and don't replace them. Consume far less.
8. Avoid purchasing and using plastic and single-use items.
9. Make your home energy efficient. If possible, move off-grid.
10. Never drive above the speed limit, and check tires for proper inflation to save gas.
11. Take a reusable container to pick up take-out pizza and coffee.
12. Work four ten-hour days, rather than five eight-hour days.
13. Shop in bulk and bring containers and bags instead of using store-supplied disposables.

14. Avoid using Facebook and cloud data storage.
15. Become an activist and join the Voice of the Streets.
16. Have no or fewer children.
17. Vote progressive.

APPENDIX D

Draft Public Official Contact

To: [Recipients such as your governor, city manager, and city council]
From: [Your name, phone number and email address]
Re: Climate, Species Extinction, and Soil Loss Crises
See: https://thesolutionsproject.org
Date: [Current date]

At this juncture, there is no longer any doubt that the planet is in a crisis.

The climate crisis is the most serious challenge to face humankind. Responsible people are searching for ways to be helpful. Governments are grappling with the question of what to do too. The goal of this letter is to bring an achievable solution to your attention.

The transition out of fossil fuels and a 100 percent renewable energy future is within our grasp. Mark Jacobson, a civil and environmental engineering professor at Stanford University, has developed a plan to begin transitioning the United States from dependence on fossil fuels to 100 percent renewable energy by 2050.

The fifty-state plan is posted on the website of the Solutions Project, a nonprofit outreach effort run by Jacobson and his colleagues to raise public awareness about switching to clean energy produced by wind, water, and sunlight. The website includes an interactive map that demonstrates the renewable resource potential of all fifty states, with explanations of how each state can meet virtually all of its power demands (transportation, electricity, heating, etc.) no later than 2050 by switching to a clean technology portfolio that is, on average, approximately 55 percent solar, 35 percent wind, 5 percent geothermal, and 4 percent hydroelectric energy sources.

Energy behemoths are excluded from the plan as unnecessary. A study of the 2020 energy policy group Rewiring America found that an aggressive push toward 100 percent renewable energy would save American households as much as $321 billion nationally or up to $2,500 per household per year.

Please confirm receipt of this letter. Thank you very much for your attention.

APPENDIX E
On "The Voice of the Streets"

I HAPPENED ON THE PHRASE "the Voice of the Streets," in Rebecca Solnit's essay "The Butterfly and the Boiling Point" in *Encyclopedia of Trouble and Spaciousness*. Solnit quotes a former reformist secretary: "The government is telling us that the street is not the place for things to be solved, but I say the street was and is the place. The voice of the street must be heard." Due to its importance, I chose to capitalize the phrase. Solnit observed, "The boiling point of water is straightforward, but the boiling point of societies is mysterious." And when those in authority ignore the needs of the people, the unknowable catalyst for change often finds voice in the streets.

Although the phrase "voice of the streets" is commonly now associated with rap music, it has long occupied a place in American political discourse. The cherished idea of civil disobedience—the active refusal to obey certain laws, demands, or commands of government or an occupying power—was settled early on when Thomas Jefferson made the importance of disobedience unambiguous: "If a law is unjust, a man is not only right to disobey it, he is obligated to do so." And Henry David Thoreau said, "Anyone in a free society where the laws are unjust has an obligation to break the law." Martin Luther

King Jr. echoed this principle regarding breaking the law: "One has a moral responsibility to disobey unjust laws."

There are many hundreds of examples establishing the veracity of civil disobedience as not only morally sound but the most effective means of social change in the United States—whether it is Rosa Parks refusing to move to the back of a bus or more than a thousand climate change activists refusing to obey an order to disband protest at the White House. Yet, in the new millennium, a developing governmental response to environmental activism demonstrates how seriously corporately funded government fears civil disobedience. In 2008 climate activist Tim DeChristopher posed as a bidder at a Bureau of Land Management auction and outbid fossil fuel interests, wreaking havoc over the land-auction process. Even though the bidding process was later deemed improper and the leases nullified, DeChristopher was prosecuted, convicted, and spent twenty-one months in federal prison for his nonviolent effort to prevent oil and gas drilling on thousands of acres of public land. In spite of the fact that funds were raised to complete the purchase based on his bid, the prosecution persisted.

In 2005 the FBI declared animal rights activists to be the nation's number one domestic terrorism threat, and a year later the U.S. Congress passed the Animal Enterprise Terrorism Act. The act was written and paid for by the agricultural and pharmaceutical industries, and young animal rights activists now go to federal prison with felony convictions on their records for freeing caged animals used for testing and living in cruel conditions. The Voice of the Streets is also under assault with proposed legislation designed to deter dissent: prison sentences for protesters blocking traffic, criminalizing the use of costumes during protests, indemnifying drivers who injure protesters, and automatically expelling students who violate protest laws. These laws are similar to those that have been used to limit or outlaw labor unions organizing and civil rights groups' protests and to restrict the rights of assembly. Since the election of President Donald Trump, anti-protest legislation has been proposed in more than twenty states. A case of content-based discrimination against freedom of speech and assembly and selective prosecution occurred when Washington, D.C., police "corralled" and arrested more than two hundred people demonstrating at President Trump's inauguration and collectively charged them with felony rioting, felony incitement to riot, conspiracy to riot, and five property damage crimes. They each faced more than sixty years in prison.

The first group of six defendants was acquitted by a jury in December 2017 in a rebuke to the United States Attorney's Office for the District of Columbia. The prosecution thereafter dismissed charges against an additional 129 defendants, leaving fifty-nine defendants still facing prosecution. One of the jurors from the first trial described their decision-making process. After they returned "not guilty" verdicts on all charges, the juror, identified only as Steve, told the media collective *Unicorn Riot* for a piece published December 21, 2017,

> It was not a close call. The prosecution admitted the morning of day one that they would present no evidence that any of the defendants committed any acts of violence or any vandalism. From that point, before the defense ever uttered a sound, it was clear to me that ultimately, we would find everyone not guilty. And while there was a great deal of careful discussion among the jurors, at no point . . . did it seem even possible that a guilty verdict would come down. This was not close.

After the initial trial ended in not guilty verdicts, not only did the prosecution dismiss charges against more than a hundred defendants, but it also eventually dismissed charges against ten more defendants and finally dismissed all charges against the final defendants.

ABOUT THE WOOD ENGRAVINGS

ALL WOOD ENGRAVINGS WERE commissioned for this book and rendered by Silver City, New Mexico, wood engraver David Wait. David and his wife, Charmeine, live in the Burro Mountains west of Silver City, where snowmelt and rainwater flow into the forty-eight-mile-long Burro Ciénaga watercourse that runs through the Pitchfork Ranch. The engravings were originally printed on David's handmade wooden Gutenberg-style press.

INDEX

Note: Page numbers in *italics* denote figures.

ABOUT THE AUTHOR

After thirty-two years as a small-town general practicing lawyer in Casa Grande, Arizona—in which A.T. Cole successfully defended a dozen murder cases, two of which risked the death penalty, served as co-counsel in the largest personal injury jury verdict in Arizona history, and served as chair of the Arizona Humanities Commission—Cole retired with his wife, Lucinda, to the ranch in southwest New Mexico. Their focus is restoration of a ciénaga, habitat for wildlife, and thwarting the climate and companion crises. They hope to draw down the carbon they spewed during their lives and encourage others to do likewise. A.T. can be reached at atandcinda@gmail.com.